A Textbook on
Advances in Production Technology of
Temperate Vegetable Crops

About the Editors

Mr. Amit Kumar is currently pursuing his Ph.D. in Horticulture (Fruit Science) from CoH, SVPUA&T, Meerut, Uttar Pradesh. He has successfully qualified for ICAR-ASRB NET in Fruit Science in 2021. He completed his B.Sc. (Hons.) in Horticulture in 2019 from CoH, BUA&T, Banda, Uttar Pradesh, and his M.Sc. (Ag) in Fruit Science and Horticulture Technology in 2021 from CoA, OUA&T, Bhubaneswar, Odisha. Mr. Kumar was selected for the award of the ICAR PG Scholarship. He was also awarded the Best Researcher Award (2021), Best Master Thesis Award (2022), Best Research Scholar Award (2022), and Young Horticulturist Award (2023) by various Professional Societies. He has published 16 Research Papers, 16 Review Papers, 35 Popular Articles, 19 Book Chapters, 03 Books, 01 Practical Manual, & More than 30 Abstracts as well during his academic years. He has actively participated and presented his work in various National/ International Conferences, Training, Seminars, Webinars, and Workshops.

Dr. Satya Prakash currently serves as a Professor & Head of the Department of Fruit Science at the College of Horticulture, Sardar Vallabhbhai Patel University of Agriculture & Technology in Meerut, Uttar Pradesh. Previously, he served as the Senior Scientist and Head at K.V.K. Muzaffarnagar from December 2003 to May 2013. There he promoted Floriculture and Vegetable intercropping with sugarcane and Gerbera cultivation in ventilated poly house increasing the farmer's income and maximum utilization of natural resources. In recognition of his work, K.V.K. Muzaffarnagar received the "Best K.V.K. Zonal Award 2013" from ICAR, New Delhi. From May 2013 to April 2018, Dr. Prakash served as the Professor and Head at K.V.K. Saharanpur. There he promoted Export quality production of Horticultural crops and Mushrooms. Again, based on his contributions, K.V.K. Saharanpur was honored with the "Pandit Deendayal Krishi Vigyan Protsahan Award 2016" from ICAR, New Delhi. Throughout his career, Dr. Prakash has received awards and appreciation from various organizations, including Managing Directors U.P.B.S.N, World Bank Mission, District Magistrate, and Ministry of Agriculture, GOI. He has been honored with the SCSI Gold Medal award, Leadership Award, Summer Memorial Award, Young Scientist Award, R.S. Yadav Outstanding Award, Fellow Award, Outstanding Horticulturist Award, and Distinguished Scientist Awards from several reputable societies. Dr. Prakash has also made significant contributions to academia, having guided three Ph.D. and seven postgraduate students. He has authored 110 research papers in NAAS-rated journals, eight books, 26 book chapters, nine technical bulletins and manuals, 81 articles, and presented 105 abstracts. Moreover, he has presented his research papers at 15 international conferences and 39 national conferences. Dr. Prakash has delivered 135 horticultural talks on ETV U.P, four talks on Doordarshan Delhi, and 158 talks on CRS. He has also participated in training and study missions in Israel, Egypt, Indonesia, and Nepal, where he showcased his work in international forums. Furthermore, he has represented his work in international forums through training and study missions in Israel, Egypt, Indonesia, and Nepal.

Mohd Wamiq is currently pursuing his Ph.D. in Horticulture in the Department of Vegetable Science at the College of Horticulture, Sardar Vallabhbhai Patel University of Agriculture and Technology, Meerut, Uttar Pradesh. He obtained his undergraduate and post-graduate degrees from the Sam Higginbottom University of Agriculture, Technology and Sciences in Prayagraj, Uttar Pradesh in 2020. In 2021, Mr. Wamiq successfully qualified for the ICAR ASRB-NET in Vegetable Science. Mr. Wamiq has published several research papers and review papers in both national and international journals. Additionally, he has authored several articles in various magazines.

Mr. Chhail Bihari completed his B.Sc. (Hons) in Horticulture in 2019 from CoH, BUA&T, Banda, Uttar Pradesh, and his M.Sc. (Ag.) in Vegetable Science in 2021 from DRPCAU, Pusa, Samastipur, Bihar. Currently, he is pursuing his Ph.D. in Horticulture (Vegetable Science) at the College of Horticulture, School of Agricultural Sciences, Nagaland Central University, Medziphema, Nagaland. He successfully qualified for the ICAR-ASRB NET in Vegetable Science in 2021 and achieved the 23rd rank in JRF conducted by ICAR AIEEA (PG). He was awarded the ICAR-JRF and UGC-Non-NET fellowship during his Ph.D. studies. In addition, he received the Best Master's Thesis Award and the Best Research Scholar Award in 2023 from various professional societies. He has published a few research papers, articles, and book chapters and presented his work at various national and international conferences and seminars.

Mr. Sajeel Ahamad completed his B.Sc. (Hon's) Horticulture with a Gold Medal in 2019 from CoH, BUA&T, Banda, Uttar Pradesh and M.Sc. (Ag.) Postharvest Technology in 2021 from ICAR-IARI New Delhi. Presently he is pursuing his Ph.D. in Horticulture (Postharvest Management) from ICAR-IARI New Delhi. He qualified ICAR-ASRB NET in Fruit Science and Vegetable Science in 2021, and 2023 respectively. He has secured AIR 8th rank in JRF and 1st rank in SRF conducted by ICAR AIEEA (PG), and ICAR AICE-JRF/SRF (Ph.D.) in 2019 and 2021 respectively. He was selected for the award of ICAR-JRF and SRF fellowship. He was also awarded the Best Master Thesis award, Best Research Scholar Award, and Young Horticulturist Award by various professional societies. He has published 7 review papers, 10 popular articles, 12 book chapters, 2 Books, and 15 abstracts and presented his work in various national and international conferences and seminars.

A Textbook on Advances in Production Technology of Temperate Vegetable Crops

Editors

Mr. Amit Kumar
Ph.D. Research Scholar
Department of Fruit Science, College of Horticulture
Sardar Vallabhbhai Patel University of Agriculture and Technology
Meerut-250110, Uttar Pradesh, India

Dr. Satya Prakash
Professor & Head
Department of Fruit Science, College of Horticulture,
Sardar Vallabhbhai Patel University of Agriculture and Technology
Meerut-250110, Uttar Pradesh, India

Mr. Mohd Wamiq
Ph.D. Research Scholar
Department of Vegetable Science, College of Horticulture,
Sardar Vallabhbhai Patel University of Agriculture and Technology
Meerut-250110, Uttar Pradesh, India

Mr. Chhail Bihari
Ph.D. Research Scholar
Department of Horticulture (Vegetable Science)
Nagaland Central University, SAS, Medziphema-797106, Nagaland

Mr. Sajeel Ahamad
Ph.D. Research Scholar
Division of Food Science & Post-harvest Technology
Indian Agricultural Research Institute, New Delhi-110012, India

PMW

PMW

FU 75, Pitam Pura
New Delhi 110034
publishmywork23@gmail.com
Mob. 91-8447075807

ISBN: 978-81-96458-83-6

MAHATMA GANDHI UNIVERSITY OF HORTICULTURE & FORESTRY
CAMP OFFICE: KRISHAK SABHAGAR, RAIPUR 492 012 (C.G.) INDIA
mguvv.ac.in vcmguvv@gmail.com (O) 0771-2991101

No./VC-Cell /2023/ Dated 17/08/2023

Dr R S Kureel
Vice Chancellor

FOREWORD

Temperate vegetables hold immense importance in our diets due to their nutritional value, versatility, and impact on sustainable food practices. These cool-season vegetables are not only delicious but also offer a plethora of health benefits. Packed with essential vitamins, minerals, and fibre, they contribute to a balanced and healthy diet. These vegetables are rich in antioxidants, support our immune system and reduce the risk of chronic diseases. With their low caloric and fat content, temperate vegetables are beneficial for health management. Additionally, their adaptability to various climates allows for year-round availability, promoting sustainable food practices. Including temperate vegetables in our meals ensures we receive essential nutrients while enjoying delicious and diverse culinary options. Furthermore, as more emphasis is placed on sustainable agriculture, temperate vegetables are a viable solution with their low water requirements and ability to improve soil health. The future of temperate vegetables looks bright, as they play a crucial role in meeting the growing demand for nutritious and environmentally-friendly food options. India, the second-largest vegetable producer after China, has been continuously expanding its vegetable production over the past few years. The present total production of vegetables in India is 200.44 MMT and the area is 10.85 million hectares as per NHB 2020-21, and the demand for temperate vegetables has also increased significantly.

It is my pleasure to introduce this valuable **"A Textbook on Advances Production Technology of Temperate Vegetable Crops"**. This book covers a wide range of aspects related to temperate vegetables, including their introduction, origin, and distribution, importance and uses. nutrient composition, taxonomy, and botany, followed by soil and climate, field preparation, commercial varieties, seed rate, nursery management, sowing and transplanting, nutrition and water management, use of growth regulators, intercultural operations and physiological disorders. It also covers the latest data on the area and production of temperate vegetables, as published by the National Horticulture Board (NHB) in 2021. In addition, the book also provides comprehensive information on the management of major insect pests and diseases, harvesting, and post-harvest management. It also covers yield, future prospects, and references related to temperate vegetable production.

The authors of this book have done an excellent job of compiling all the necessary information related to temperate vegetable production technology. The book is written in simple language and is easy to understand for students, teachers, researchers, and innovative vegetable growers.

I am confident that this book will serve as a valuable resource for all those involved in the production of temperate vegetables, and it will contribute significantly to the development of the vegetable industry in India. I congratulate the authors for their excellent work and thank them for sharing their knowledge with us.

Place Raipur
Date 17/08/2023

(R. S. Kureel)
Vice-Chancellor

Formerly:

- **Birsa Agricultural University, Kanke, Ranchi, Jharkhand**
- **Dr. B.R. Ambedkar University of Social Sciences, MHOW, Indore, M.P.**
- **Acharya N.D. University of Agriculture & Technology, Ayodhya, U.P.**

Preface

Vegetables, specifically, hold a crucial position within horticulture, playing a significant role in driving our economy forward. In recent years, the demand for temperate vegetables has increased significantly, leading to the need for advancements in production technology to meet this demand. This has resulted in the development of innovative techniques for fruit cultivation. The advances in production technology have not only increased the quality and quantity of temperate vegetables but also led to higher yields and improved economic returns for growers. Therefore, it is crucial to continue to invest in the development and implementation of advances in production technology to meet the increasing demand for fruits and ensure a sustainable and prosperous future for the vegetable industry.

This comprehensive guide provides in-depth information on all aspects of temperate vegetable production. This book satisfies the demand for a quality reference cum edited book among students. The authors have done an exceptional job of compiling all the necessary information related to temperate vegetable production technology. They have presented it clearly and concisely, making it easy to understand and apply. Every effort has been taken to assure the accuracy of the data gathered from a variety of sources, including people, organizations, institutions, and publications. We expect that this book will be a helpful resource for educators, professionals, and learners seeking employment in ICAR institutions, State Agriculture University, State Public Services/UPSC/ARS/JRF/SRF/ASRB-NET *al*ong with entrance examinations for B.Sc., M.Sc., and Ph.D. Degree in Vegetable Science and also very useful for progressive farmers, and horticultural extension workers, and boost farmers' income.

We are thankful to our all authors, contributors, friends, seniors, and juniors for their excellent work, and thank them for sharing their knowledge with us and for their moral support. We are also thankful to Publication to bring the finest format of this book.

We are confident that readers will greatly benefit from the content of this book in their various pursuits. We wholeheartedly welcome any suggestions or ideas to enhance its educational value. We strive to make this book an even more enriching experience for readers by incorporating educational features that promote deeper understanding and practical application of the concepts discussed. Your input and contributions are highly appreciated in this endeavor.

10th July 2023 **Editors**

Contents

Root Crops

Pea and Beans

Leafy Vegetables

Salad Vegetable

Perennial Vegetable

Potato

1

Advances in Production Technology of Cabbage

***Suprava Biswal*[1*]*, Toko Yarin*[2]* and Raj Kumar*[3]**

[1&3] *Ph.D. Research Scholar, Department of Vegetable and Spice Crops, Uttar Banga Krishi Viswavidyalaya, Pundibari, Cooch Behar-736165, West Bengal*

[2] *Ph.D. Research Scholar, Department of Plantation Crops and Processing, Uttar Banga Krishi Viswavidyalaya, Pundibari, Cooch Behar-736165, West Bengal*

**Email: supravabiswalchc2@gmail.com*

Introduction

Cabbage (*Brassica oleracea* var. *capitata*) belongs to the family Cruciferae or Brassicaceae having chromosome number 2n=18. 'Cole wart,' a wild non-heading variety of *Brassica oleracea* var. *sylvestris*, is the ancestor of cabbage. The majority of the 100 species that make up the genus Brassica are indigenous to the Mediterranean region. It is also believed that Germany is where the true-headed cabbage originated. Currently, mostly cabbage is grown in the country like Caribbean, Indonesia, Malaysia, Central Asia, and Western and Eastern America.

Figure 1.1: Cabbage

The third-largest producer of cabbage worldwide is India. The majority of the nation's states where it is grown include UP, Orissa, Bihar, Assam, WB, MH, and Karnataka. Annually in India, cabbage was produced around in the area of 418 thousand hectares and production of 9.72 million metric tons with the productivity of 22.7MT/hectares (NHB, 2021-22). The other main producers in the globe, in addition to India, are China, Russia, Japan, Korea, the USA, Indonesia, and Ukraine.

Economic Importance and Uses

Cabbage is mostly used as vegetable and has many nutritional properties. The glycoside sinigrin is responsible for the flavor of cabbage leaves. Goitrogens, which are found in cabbage, make the thyroid gland expand. The thio glucosides (sinigrin), which depend on enzymatic breakdown in the presence of the enzymes (progoitrin and epi-goiter), are non-goitrogenic and serve as the building blocks for the anti-thyroid molecule known as goiter. Following hydrolysis, the goiters is created by cyclizing an unstable isothiocyanate that contains a hydroxyl group. White cabbage cultivars' total glucosinolate concentrations ranged from 299 to 1288 ppm, while their goiters concentrations ranged from 1.2 to 26.0 ppm (Swarup, 2006). Even though the glucosinolates' enzymatic hydrolytic products are goitrogenic, some of their anticancerous properties exist in breakdown products. Additionally, it has been suggested that boiling cabbage can cause the majority of its goitrogen characteristics to be lost (Anonymous, 2012). Cabbage leaves are recommended for cough, fever, skin conditions, peptic ulcers, urinary retention, andhemorrhoidss in ayurvedic medicine. Concentrated cabbage juice is highly beneficial at treating peptic ulcers, according to a clinical trial. Cabbage is an excellent source of a sulphur-containing amino acid. Fresh cabbage juice contains antibacterial capabilities and has been shown to inhibit the growth of certain lactic acid bacteria strains.

Cabbage is a good source of sulphur containing necessary amino acids. Carotene (provitamin A), ascorbic acid, riboflavin, niacin, and thiamine are present in significant levels. The amount of ascorbic acid in 100 g of fresh weight fluctuates between 30-65 mg (Bose *et al.*, 2002). When cabbage is cooked or shredded and exposed to the air, ascorbic acid can be significantly lost. An excellent source of anthocyanin pigments is red cabbage. Red cabbage contains both mono and diacylated cyanidin compounds. The thermo labile sulphur compounds in cabbage produce sulphurous odours. Dimethyl disulfide and hydrogen sulphide, two of cooked cabbage's main volatile sulphur compounds, are formed from a free amino acid known as 1, S methyl cysteine sulfoxide. There have also been reports of 2-propenyl isothiocyanate and 3-butenyl isothiocyanate being present in red and white cabbage.

Botany

Most of vegetable crops are raised from seeds. Cabbage is one of the important cold season vegetable crops also raised from seed as like another cruciferous crop. Cabbage seedlings has cordate shaped cotyledonary growth with a thin tap root system. The crop shows vegetative growth in the first season whereas for seed production and flowering occurs in thesecond season. At the mature vegetative stage, plant height varies from 50-60 cm but it can grow up to height of 150-200 cm at the flowering time in the 2nd year of growth. There are 3 different shape of cabbage head available according to days taken for maturation like early maturing variety have small size with round in shape whereas late maturing variety have flat head shape with bigger size therefore average weight of head varies from 500 gm-3kg. Cruciferous crop has peculiar leaf characteristic like waxy leaves helps in reduction of transpiration and respiration losses, thick and lobed or highly dissected leaf margin. Cabbage is a shallow and fibrous rooted crop having more root mass observe in the 25-30 cm depth of the soil.

The cabbage inflorescence is an indeterminant, unbranched terminal raceme type and yellow in color measuring about 50-90 cm in height. Each flower of inflorescence has four petal, four sepal, six stamen of tetradynamous condition that is four long stamen and 2 short stamen and a superior ovary. It is highly cross-pollinated crop so 95% of pollination is carried out by insects. The fruit of the cabbage known as silique type which is dehiscent type that pod opens at maturity time. Self-incompatibility is strong in cabbage crop as compared to other cruciferous family crop.

Classification

Cabbage is also classified into 7 divisions was suggested by Bailey (1930)

A. Open foliage type

i) Branching shrubby kales - *Brassica oleracea* var. *fruticose*

ii) Common biennial mostly unbranching kales- *Brassica oleracea* var. *acephala.*

iii) Low cabbage like Portuguese kales- *Brassica oleracea* var. *tronchuda*

B. Compact leaf head or bud type

iv) Edible terminal head, the cabbage- *Brassica oleracea* var. *capitata.*

v) Edible lateral small heads or buds, the Brussels sprouts- *Brassica oleracea* var. *gemmifera.*

C. Inflorescence grown as modified thickened flower shoots of malformedflowers

vi) Flowers only partially if at all modified- *Brassica oleracea* var. *italica.*

vii) Flowers thickened and malformed into a head or several head- *Brassica oleracea* var. *botrytis.*

Classification of Variety

1. Classification based on origin

a) *Brassica oleracea* var. *capitata* ssp. *Mediterranea* (Mediterranean cabbage)

b) Brassica oleracea var. capitata ssp. Orientalis (Oriental cabbage)
c) *Brassica oleracea* var. *capitata* ssp. *Europea* (European cabbage).

2. Classification based on color of heads

a) *Brassica oleracea* var. *capitata* L.f. *alba* (white cabbage)
b) *Brassica oleracea* var. *capitata* L.f. *rubra* (red cabbage)
c) *Brassica oleracea* var. *sabauda* (savoy cabbage)

3. Classification based on head shape

a) Round shape head or ballhead type (Golden Acre, Pride of India and Copenhagen Market)
b) Flat head or drum head type- Flattened head from both ends (Pusa Drum Head)
c) Conical head type- Heads are small and very early type (Jersy Wakefield, Charleston Wakefield)
d) Savoy type- Leaves are wrinkled type and high-quality heads (Drumhead Savoy, Chieftain)

4. Classification based on size and shape of the head (Alam, 1936)

a) Danish ball head group- thin leaves, compact solid heads, medium sized, good keeping quality, fine texture (Danish Ball Head)
b) Alpha group-The earliest grouphead, very small, solid head limited commercial uses, (Miniature Marrow)
c) Volgagroup-thick leaves, shining blue colored leaves, loose bottom (Volga)
d) Red cabbage group- similar to Danish Ball Head except red color of leaves (Red Rock, Red Dutch, Large Red).

Commercial varieties

Bajrang (BSS-50)

This is a private sector hybrid variety which is highly tolerant to heat stress and resistant to fusarium wilt released at Beejo Sheetal Seeds company. It takes 65 DAT for head maturation buthas longer staying capacity in the field up to 45 days after maturity appears of the head and weight of the head is around 800 gm^{-1} kg.

BSS-44

This is another private sector released hybrid developed at Beejo Sheetal Seeds company. Round head with average weight is 2.5 kg. Mature in 75 days after transplanting. This variety has unique characteristics and can stand in the field for 20 days without affecting its marketable quality. Resistant to fusarium wilt.

Copenhagen marke

Introduced cultivar, highly popular among farmers due to the early in nature of the small size head. Olaniyi and Ojetayo (2011) used different types of organic fertilizersr like alesinloye organomineral fertilizer, sunshine organomineral fertilizer, neem compost, cassava peel compost, pacesetter organomineral fertilizer on Copenhagen market and F_1 milor cultivar of cabbage. They concluded that with or without fertilizer Copenhagen market gave the better result as compared to F_1 Milor variety.

Early Drumhead

Early type cultivar with flat head shape popular among farmers.

Golden Acre

Earliest type cultivar suitable in hilly place during spring-summer time usually takes about 60-65 DAT for harvesting of the head. Selected from Copenhagen market. Solid heads and weighing of about 1.0-1.5 kg. Cracking is a main demerit of this cultivar so to cope up with this issue, harvesting can be done immediately after head formation. Conical heads may produceif timely sowing is not done.

H-113

IARI, Katrain, self-incompatible hybrid developed from crossing between two lines i.e., Sel-8 and 83-6. Head is round in shape with bluish and waxy leaves. Suitable for harvesting at 60-68 DAT with average weight of the head is 473.5 q /ha.

Kranti

This is an early type cutivar takes about 60 DAT for full head formation. Suitable for close spacing. Round head shape with 800gm-1200gm weight of each head.

Jersey Wakefield

Introduced variety having pointed head shape. Average weight of 1-1.5 kg/ head. Less outer leaves with usually takes 55-60 DAT for attaining marketable stage of the head.

Late Drumhead

105-110 DAT for maturation of head. Smaller head with short stalk and contains less leaves. Uniform heads, solid, flat type head shape.

Pride of India

It is popularized in Dr. YS Parmer University, Solan. Average weight of the head is 1.5-2 kg.

Pusa Ageti

F1 hybrid produced through simple recurrent selection technique from Taiwan. Seed production is possible in sub-tropical climate. Average head weight is 600-1200 gm with yield is 110-380 q/ha. Takes about 70-90 DAT for maturation of head.

Pusa Drum Head

This a black leg resistant cultivar (Swarup *et al.*, 1968) developed at IARI, Katrain. Head is flat in shape with one to four kg weigh. Takes about 80-90 DAT for head maturation.

Pusm Mukta (Sel.8)

Hybridization between EC 24855 x EC 10109 developed at IARI, Katrain. The key character of this cultivar is light green foliage with wavy margins and resistant to black rot caused by bthe acteria *Xanthomonas campestris*. Slightly flat head shape with head weight of 1500 gm to 2000 gm.

Pusa Synthetic

Synthetic cultivar, early type in nature developed at IARI, Katrain. Average yield potential is 38-40 ton/ha.

Red Cabbage

All the red type of cabbage cultivar has the tolerant capacity against DBM. It usually takes 90 DAT for head formation with average weight 1-2 kg.

September

Introduction from Germany but most popular in the Nilgiris hills, TN. Solid, flat and round head type. It usually takes about 96-100 DAT for maturity appears of the head.

Sri Ganesh Gol

Hybrid cultivar mostly resistant to yellowing diseases developed at Mahyco, private seed company, MH. It is highly recommended by farmer due to good transport quality and preferred in the eastern and southern part of India. Maturity appears after 90-95 DAS. Average yield of this cultivar is70-80ton/ha.

Suvarna (BSS-32)

This is a private sector released hybrid cultivar matures in 100-110 days and developed at Beejo Sheetal Seeds company. Head is round in shape with 3000gm average weight of the head.

Questo

It is a highly resistant to heat stress condition and can stand in the field up to 70 days without deteriorating its head quality.

Pusa Cabbage-1

High temperature tolerant hybrid variety having compact, smooth early type head mature around 65 DAT with an average weight of the head is 800-1200 gm. This variety has longer field staying capacity that is around 45 DAM (days after maturity). It is highly resistant to *Fusarium* wilt also.

Kinner Red

This variety has red color head and all the red color cabbage cultivar having distinct characteristic of waxy coated leaves which helps in avoidance of DBM (diamond back moth) attack. It takes around 90 DAT for maturation with an average weight of the head is 1000-2000 g.

Climate

Cabbage can be cultivated in a variety of environmental circumstances, but a cool, humid climate is ideal. For seed germination, the ideal soil temperature is 23-25.2°C. The ideal temperature for growth ranges from 24.2 to 33.2°C. Most cultivars stop growing when the temperature rises above 42.3°C. It has been observed that after a cold treatment at 7 to 10°C for 6-9 weeks, the plant bolts early and opening of flower occurs when exposed to 7-10°C temperature for a short period of time; however, plants must have passed the juvenile phase prior to this temperature treatment (1969, Nieuwhoff).

Now-a-days increased temperature is a main reason for lowering of crop quality. Therefore, to cope up this issue modern techniques plays a significant role. Spraying of some foliar spray (brassinosteroid, methyl jasmonate, salicylic acid and sodium nitroprusside) helps in improving of antioxidant and enzymatic activity (catalase and peroxidase) under heat stress condition with regulating photosynthetic activity inside the crop, which ultimately gives the best quality of kimchi cabbage leaf (Lee *et al.*, 2019). Increased temperature (25-25°C) is also responsible for severe disease incidence in cabbage leaf like soft rot of cabbage caused by *Erwinia carotovora* sub sps. *carotovora* (Bhat *et al.,* 2010).

Soil requirement

Cabbage may be cultivated in a variety of soil types, from sandy soil to clayey type of soil. Early maturity cultivars give optimum yield in sandy to loamy type soils, but developing cultivars thrive in clayey soils. However, organic matter rich, micro pores rich soils produce higher yields. Soil pH should be between 6.0 and 6.5 for cabbage cultivation. The majority of cabbage is salt tolerant. Die back disease in leaf surface and margins of the leaf observed in saline soils, making them more prone to illnesses like black leg. White cabbage and savoy type has highest salt index denoted as C_6 whereas, red cabbage is more sensitive to salt level of the soil with denoted as C_4. At C_8 salt index, the crop may have reduction of yield around at 40%. (Bose *et al.*, 2002)

Sowing Time

The timing of seed sowing varies by region in India. Seeds are sown on hills, particularly high hills, in May - June for a summer / autumn yield. The summer and autumn crop are very limited in mountainous locations that receive strong rains, and seeding is done in autumn to harvest them in late spring or early summer by overwatering them. Sowing in situ begins on the northern Indian plains in the 1st week of Aug and continues until Nov month for late type maturity cultivars. Sowing season in the eastern part of India typically begins in 15th to end of the Sept. Sowing is postponed further in the south part of the hilly areas, with the exception of the western and southern peninsulars, where the crop can be cultivatedvirtually all year by selecting the appropriate cultivar.

Seed Rate, Method of Sowing and Transplanting Distance

It is necessary to plant 200-500 gm of seed per hectare. Cabbage is a crop that has been transplanted. Early maturity group cultivar seed beds should be covered by polythene to avoid damage against heavy rain and cold wind however in the hilly region, seedlings can be raised under glass fires or poly tunnels. Seedlings of midseason or late cultivars, on the other hand, can be grown in the open. Nursery bed should be free from pathogenic bacteria and fungal infections.In the seedbed, add 2-3 kg/m^2 sieved well rotting farmyard manure or compost. Before seed sowing in the nursery bed, seeds should be treated with various type of fungicides namely thiram, captan or bavistan @ 2 gm/kg of seeds. In the nursery bed, the optimal spacing between R-R is 10 cm whereas depth of the seed sowing should be four times from the size of the seed i.e., 1.5 to 2.5 cm. A seedbed of 200-300 m^2 at a rate of 1 to 2g/m^2 is required for planting 1ha. A one-gram seed would produce approximately 1000 number of seedlings. After sowing of seed, seeds should be covered with finely manure and soil combination. The nursery beds are slightly levelled. Availability of adequate moisture supply will be helpful for optimum seed germination and proper growth of seedlings. Another factor is mulching which is highly beneficial for keeping the nursery bed weed free with maintaining optimum moisture level. Watering is mostly supply through water cane and mulching is done by using thin layer of dry grass. After seedlings emergence, dry grass mulching material should be removed from the nursery bed to avoid lanky growth of the newly emerged seedlings. If seedlings are overcrowded as a result of thick seeding, the surplus seedlings should be trimmed off. "Damping off" caused by several fungi such as *Pythium spp.* and *Rhizoctonia spp.* is a very serious disease at the initial stage of seedling emergence stage. This condition is well treated with a 0.3% blitox spray. In general, seedlings of 5-6 weeks old are appropriate for transplantation, although seedlings up to 7 weeks old can also be transplanted. Older seedlings, on the other hand, have a negative impact on the proper establishment of root system and growth of young age plants. The best time to transplant varies according on the climate. Planting in India's plains is possible from the month of Aug to Dec, and in the hills from Apr to Aug. Early typevariety like "Golden Acre" perform well in the plains due to shorter cool periods, while "Pusa Drum Head" and other late maturing kinds perform better in the hills due to longer chill periods. The planting distance varies depending on the cultivar, planting season, and soil conditions. On the basis of cultivar maturity, the following distances are commonly recommended: early type 45 (R-R) x 45 cm (P-P) or 60 x 30 cm in a hybrid variety, mid maturity group 60 (R-R) x 45 cm (P-P), and late 60 (R-R) x 60 cm (P-P). Depending on the temperature and soil conditions, transplanting is done on flat terrain, ridges, or in furrows. The ridge approach is better suited for early planting, especially in places where rain falls at the time of planting. Transplanting should be done in the furrows on saline soils conditions.

Early type cutivar of cabbage is Brunwick is found to be best in head yield, head equatorial polar diameter ratio, head density, head volume and dry matter content at 1st Nov date of sowing with 70cm plant spacing (Abed *et al.*, 2015).

Vegetable Based Cropping System

The main objective of any cropping system is to achieve sustainable yield. Guvenc and Yildirim (2004) concluded that cabbage based intercropping system gave highest yield with high net income. In intercropping system cabbage was taken as main crop whereas intercrops were leafy type lettuce, onion, radish and snap bean. Highest parameter like yield, growth of cabbage, nutritional content was observed in cabbage. LER (land equivalent ratio) was also more than 1 in cabbage based intercropping system. But when cabbage was grown as an intercropping crop with a main crop of onion and garlic, pest population was lower in onion, garlic-cabbage cropping system recorded by Debra and Misheck (2014). Different type of infestation can also be managed through intercropping based farming system. Madriaga (1994) found that reduction of diamond back moth population in intercropped farming system (tomato, sweet pepper, cowpea and peanut) as compared to monoculture cabbage farming system. However, other pest population like fruit borer, aphid, whitefly could not able to control in intercropping system. Therefore, use of integrated pest management with suitable cropping system may be responsible to achieve sustainable yield.

Nutrient Management

Manure and fertilizer requirements in cabbage are established by soil testing and are dependent on the fertility status of the soil. If soil testing is not practicable, farmyard manure at a rate of 20-25 ton/ha should be used. It should be properly blended into the soil before transplanting within 15-20 days. Recommended dose of NPK is 150:60:60 (Sarkar and Rakshit, 2021). At the time of transplanting, half of the nitrogen and the full amounts of phosphorus and potash are applied. The remaining nitrogen is sprayed 30-45 days after transplanting. (Chaudhary *et al.,* 2015) showed that treatment consisted of 100% RDN + vermicompost @ 10t/ha. + *Azotobactor* + phosphorous solubilizing bacteria obtained higher yield of cabbage head.Nitrogen had a considerable impact on all growth and yield parameters, which grew gradually with increasing nitrogen levels up to 200 kg N/ha.

Not only inorganic fertilizer is responsible for head yield but also organic fertilizer has tremendous effect on head yield and as well helps in improving soil physical, chemical and biological property. (Goswami *et al.*, 2017) converted water hyacinth into usable compost known as drum compost mixed with locally prepared vermicompost helped in improvement of soil organic carbon content, increasing humic acid, fulvic acid content, physical stability and microbial diversity of the arable soil in cabbage farming systems. Particular mixed compost also stimulated cabbage head yield, quality and increasing storage life of the crop. Liquid foliar spray of organic component (urea @ 1g/L, humic acid @ 5 Ml/L, Higro Amin-L 5 ml/L, and Maxinim @ 5 ml/L has certain advantages in terms of improving cabbage head growth and development (Manea, 2017).

Micronutrient is playing a significant role to carried out proper growth and development of the crop. Without micronutrient, crop cannot complete their life cycle although it is required only in trace amount. (Habiba *et al.*, 2009) applied sulphur and boron during the growth period of cabbage head gave optimum result in terms of yield and quality.

Combination of inorganic fertilizer (75% RDF) with bio-inoculants (*Trichoderma harzianum*, *Bacillus subtilis*, *Pseudomonas fluorescens*) applied as seedling biopriming in red cabbage crop helps in use of decreasing amount of inorganic fertilizer as well as improvement of quality head rich in protein and other biochemical property. They also concluded that application of dual inoculation gave good result as compared to single and triple inoculation of bacterial strains (Sarkar and Rakshit, 2021).

Water Management

Cabbage crop requires plentiful supply of water for complete their life cycle. 1st irrigation is known as life-saving irrigation for any crop during establishment time therefore, application of water at 10-15 days interval during winter month is found to be effective for proper head development. But sufficient moisture level should be maintained to get good yield. Furrow method of irrigation is mostly suitable in cabbage cultivation. Flooding or waterlogged condition during head development is a dangerous during head development stage because it will cause splitting of head occurs.

Radovich *et al.* (2005) observed that supplied of water at the time of head development had more important as compared to other stages of crop growth. Characters like maximum head weight, high glucose content and less sucrose content of cabbage head were recorded at supplied irrigated plot during head development stage but glycoside like high glucosinolate content were highest at non irrigated plot.

Weed Management

Regular intercultural activities are required to support healthy plant growth by aerating the root system and controlling weeds. It is necessary to have two manual weddings. Weed management in cabbage can be accomplished with herbicides such as basalin @ 0.5 kg/ha (soil incorporation).

Harvesting, Yield and Storage

Early-type cultivars cultivated in comparatively warmer temperatures generate a loose head at first, but harden as they mature. Some cultivars' heads begin to break soon after maturity. If harvesting is delayed, the quality of the head deteriorates rapidly. As a result, harvesting should be done at the appropriate stage in order to obtain high-quality head. After transplanting, early cultivars take 60-80 days, medium cultivars 80-100 days, and late cultivars 100-130 days to harvest. Early cabbage varieties yield 33-44 ton/ha, while mid maturity and late type maturity cabbage yield 44-66 ton/ha in the northern plain. Cabbage may be preserved for 2 to 8 months at 0°C and 90-95% RH (Murugiah *et al.*, 1987).

Blanching

Blanching is a heat treatment required before pre-storage period with an aim of lowering the enzymatic activity to avoid storage loss of cabbage head. Sarkar *et al.* (2021) conducted an experiment by heat treatment of cabbage head at 80°C, 90°C and 100°C temperature for 12, 8 and 2 minutes, respectively. They found that after blanching treatment at 100°C temperature with 2 minutes helped in retention of phenolic compound and greenness of cabbage head.

Major Diseases and Pests and their management

Diseases

i) Fungal diseases

a) **Downy mildew** - causal organism i.e., *Perenospora parasitica*

Symptoms - Plants can become infected at any stage during their growth cycle. Young plants are highly susceptible show white mildew growth on the underside of the leaf.Later on, white mildew growth enlarged and become yellowish color on the upper side of the leaves. A juvenile leaf may fall off. Older leaves normally survive, and infected regions grow in size as they turn scorching in appearance. When the disease becomes severe, the entire leaf dies and affects the photosynthesis system in the plant and ultimately plant dies.

Control measures - The effective control measure of this downy mildew growth can be spray with copper oxychloride @ 0.3% at the seedlings stage in the nursery bed.

b) **Black Spot** - causal organism i.e., *Alternaria brassicae*

Symptoms - The initial sign is a little dark spot on the surface of the leaf. Concentric rings form as the spot expands. The infected spot grows in size and defoliates plants. It is a seed crop disease that destroys pods and seeds.

Control measures-This type of fungal infection can be prevented by foliar spray with the captan, thiram @ 0.2% or copper oxychloride @ 0.5%.

c) **Cabbage yellow** - Causal organism i.e., *Fusarium oxysproum* f. sp *conglutinans*

Symptoms - Stunted plant growth with shrunken yellow appearance and plant growth is very limited. The fungus spreads via seed and can live for many years in soil. Warm temperature conditions encourage disease development.

Control measures - Disease management is challenging. To some extent, field sanitation, crop rotation, and the use of disease-free seeds can help control the disease. Growing the resistant or tolerant type of variety from cabbage yellow diseases. There have been reports of two types of resistance.

Type A - Resistance is controlled by a single dominant gene. It remains stable at varying soil temperatures. This appears to be a direct descendant of *Brassica oleraceae* var. *sylvestris*.

Type B - This resistance is found in the cabbage cultivar Wisconsion Hollander.

d) Black leg - Causal organism i.e., *Phoma lingam*

Symptoms - It occurs in the majority of regions, particularly in areas with high rainfall throughout the growing season. Because the fungus is transported by the seed, it might appear at any stage.When the damaged plant's stem is severed vertically, the sap stream is severely discoloured. The entire root system decays from the ground up. The afflicted plants frequently collapse in the field.

Control measures - The main source transfer of diseases through seeds can be avoided by hot water treatment of the seeds. Spraying with copper oxychloride or an organomercuric chemicalcan be help in prevention of this disease and using resistant varieties such as Pusa Drumhead will be effectible for removal of disease.

ii) Bacterial Diseases

a) Black rot of crucifers - Causal organism i.e., *Xanthomonas campestris* pv. *campestris*

Symptoms-Infected plants grow stunted, with one side of the cotyledon turning yellow to black and bending down and dropping off early. Many lower leaves fall off early as well. The surviving leaves may yellow and have dark veins. From the leaf margin, 'V' shaped chlorotic to yellow lesions develop. The veins and veinlets become dark, and the leaf tissue becomes necrotic and brittle. The lesion progresses towards the midrib, and the infection spreads farther into the stem and root as a result of this systemic infection. Vascular tissues in affected areas turn dark, followed by the internal breakdown of squishy tissue. Succulent seed stocks and siliqua may acquire black sunken patches.

Control measures - This type of malady can be prevented by spraying of plantomycin @ 100 ppm or Agrimycin @ 100 ppm.

b) **Club root -** Causal organism i.e., *Plasmodiophora brassicae*

Symptoms - Club root infection is mostly prevalent in cole crops. It is mostly seen in saline affected soil where pH is higher than 7. Affected root shows irregular galls in the tap root system whereas spindle shape in the lateral roots ultimately it affects the plant growth results wilting of the plant occurs.

Control measures - Treated with 4% calomel paste of the seedlings root at the time of transplanting time can be effective for the control of this diseases.

Major pests of cabbage

i) **Leaf webber (*Crocidolomia binotalis*)-**The eggs present in a mass and joint together by production of viscous glue type substances of this pest mostly seen under the leaf surface. Adult moths are mostly slender type and light brown in color.

Control measures

a) Most insecticides used to control diamond back moths are also effective against leaf Webber.

b) In biological control, key natural mortality causes for leaf webber were observed in late larval instars as parasitization by *Bracon hebetor*, i.e., *Palexorusta solennis* (Singh *et al.*, 1978).

ii) **Mustard sawfly (*Althalia lugens proxima*)** - In the early stages, the blackish caterpillar targets all cole crops and feeds on the leaves of immature seedlings. When contaminated leaves are touched, they curl and fall to the ground.

Control measures - Hand picking caterpillars is the greatest way of controlling the larvae if the infestation is modest and the number of attacked plants is limited.

iii) **Mustard aphids (*Lipaphis erysimi*)** - When the weather is cloudy, this little greenish white bug attacks cabbage and other cole crops. They feed on the sap of the plants. Plants wither and die as the damaged leaves curl.

Control Measures

i) Aphids is the serious pest for plant due to infection starts from the nursery bed and can be controlled effectively by spraying of rogor, metasystox etc. @ 1 ml/lt. of water starting from the nursery stage at 15 days interval.

ii) The other method can be used to control the aphid infestation that is soil incorporation of thimmet @ 12-15 kg/ha. in two separate doses like half portion of the recommended doses at planting time and rest half portion at the time of earthing up.

iv) **Diamond back moth (*Plutella xylostella* L.)** - It was originally seen on cruciferous vegetables in India in 1914. A little pale green caterpillar that eats on leaves and chews on them. The contaminated produce is no longer marketable.

Control measures

a) This pest infestation can be reduced by use of trap crop that is mustard crop and that should be planted 15 days earlier before main crop that is cabbage crop. Foliar spray twice or three times with Cypermethrin (0.2 kg a.i. /ha), Femvalerate (0.1 kg a.i./ha), or Sulprofos (1.0 kg a.i./ha).

b) It can also be managed by foliar spraying of padan @ 500gm/ha. in cabbage at 15 days interval starting from the leaf primordial stage. Other type of pesticides can also be effective to control of DBM pest by spraying of cypermethrin @ 0.2 kg/ha. or femvalerate @ 0.1kg/ha.

v) **Cabbage borer (*Hellula undalis* F.)** - A whitish brown caterpillar that feeds on stems and mines the leaves. Later, it burrows inside the cabbage stem. The damaged plants wilt, become small, or die. It can be managed by foliar spraying of padan @ 500gm/ha. in cabbage at 15 days interval starting from the leaf primordial stage.

vi) **Cabbage butter fly (*Pieris brassicae* Linn.)** - In this category pest maggots are highly voracious in nature and make a hole into the plant by affecting the young rootlets system. Afterwhile affected plants become yellow, wilted and later on death of the plants occur. It can be managed by foliar spraying of padan @ 550 gm/ha. in cabbage at 15 days interval starting from the leaf primordial stage.

Seed Production

Climatic requirements

Cabbage grows in a generally cool, humid area with moderate rainfall that falls evenly throughout the growing season. In the head stage, it can withstand frost. To bolt and commence seed stalks and flower, it requires a latent phase of cool temperature. Cool temperatures, on the other hand, are beneficial only until the stem diameter is at least 1cm. This happens in temperate climates during the winter after the first season growth. The second year promotes flowering and seed production. When the headed plants exposed to a mean temperature of about 5°C for 6-8 weeks, seed stalks are emerged. Cabbage seed production is only possible in hilly areas of India.

Types of seed production method

Cabbage can be grown as annual crop for quality head production but can be grown as biennial for seed production purpose. During the autumn, the seed crop can be left in place or transferred. The in-situ method of seed production can be used for certified seed production and later method can be used for genetically pure seed production that is nucleus seed production. The crop is allowed to overwinter and generate seed in its original location, i.e., where it was first planted in the seedling stage, using the *in-situ* method. The adult plants are uprooted during the transplanting technique. After shed down of whorl, the plants with entire stem below the cabbage head are placed in the clean fields. There are three methods available for production of cabbage seed.

1. **Stump method -** After the fully grown of cabbage head in the first year, it can be inspected in the field with the objective of getting true-to-type seeds. The inspection technique can be achieved by using different procedures like plants with abnormally head shape should be removed. Afterwards leaving the cabbage stem and outer leaves, heads are chopped using the sharp knife. "stump" refers to the plant's decapitated section. The heads are marketed, and the stumps are either left in place or replanted in the second season, in the autumn. After the dormancy is broken in the next spring, the bud sprout forms the axils of all the leaves and leaf scars.

   Advantages - It can provide additional income through the selling of cabbage heads.

   In this method of seed production crop matures 12-15 days earlier than the head intact method and seed production is quietly higher also.

   Disadvantages - Flowering shoots are very weak stem therefore, heavy staking is required during carried out intercultural operation to get quality seed.

2. **Stump with central core method-**When the crop is fully grown in the first season, the heads are inspected for true to type using this procedure. Plants with abnormally shaped heads are removed and discarded. The heads are then chopped by using sharp knife from all the sides with downward perpendicular incisions, taking care not to injure the central core. The shoots originating from the main system are not weak stem so don't require staking, which is an improvement method over the stump approach. The head is given two vertical cross cuts during the end of Feb and until the 15th of Mar, when the heads begin to burst. Avoiding damage

to the core growing point. In the absence of such cuts, the heads emerge irregularly, and the growing tip is occasionally broken. All the operation should be completed by visiting the field 2 to 3 times.

Advantages - In this method of seed production flowering shoots emerged from main stem are strong therefore seed yield is higher as compared to other method of seed production.

Disadvantages - After the chopping of cabbage head cannot be marketed.

3. **Head intact method** - After fully grown of head in the 1st season, field inspection is done at three times in the aim of get true to type seed material through removal of abnormal head shape plants. The head is left intact, with only a cross incision to allow for the formation of a stalk.

   Advantages - This type of seed production method will save time and labor.

   Disadvantages - The only disadvantage in this method is lower seed production occurs as compared to other seed production method.

Future Prospective

1. Precision agriculture: Precision agriculture technologies, such as soil sensors and remote sensing, can be used to monitor and optimize the growth of cabbage crops. By using real-time data on soil moisture, nutrient levels, and plant growth, farmers can make better decisions about irrigation, fertilization, and pest management, leading to higher yields and improved crop quality.
2. Plant breeding: Advances in plant breeding techniques, such as gene editing and marker-assisted breeding, can be used to develop new cabbage varieties with improved traits, such as higher yields, disease resistance, and better taste and texture. These techniques can also help to speed up the development of new varieties, reducing the time and resources required for traditional breeding methods.
3. Controlled environment agriculture (CEA): Cabbage can be grown in controlled environments such as greenhouses, hydroponic systems, and vertical farms. By controlling factors such as temperature, humidity, and light, farmers can optimize growing conditions and reduce the risk of pests and diseases. Additionally, CEA can allow for year-round production and reduce the environmental impact of traditional field-grown cabbage.
4. Robotics and automation: Robotics and automation technologies can be used to automate tasks such as planting, harvesting, and sorting, leading to increased efficiency and reduced labor costs. For example, automated

cabbage harvesters can increase the speed and accuracy of harvesting, reducing the amount of time and labor required.

5. Sustainable agriculture: Sustainable agricultural practices, such as reduced tillage, cover cropping, and integrated pest management, can help to reduce the environmental impact of cabbage production while also improving soil health and reducing the use of chemicals.

Overall, these advanced production technologies can help to increase the efficiency, sustainability, and profitability of cabbage production, while also improving the quality and nutritional value of the crop.

References

Abed MY, El-Said EM, Sheb EF. (2015). Effect of planting date and spacing on yield and quality of cabbage (*Brassica oleracea* var. *capitata* L.). *Journal of Plant Production.* 12(6): 2093-2102.

Abundance of diamondback moth, *Plutella xylostella* (L.) and its natural enemies in different cabbage-based cropping systems. (1994) Madriaga, Philippines University Los Banos, College, Laguna (Philippines) Thesis.

Alam, Z. (1936). Cytological studies of some Indian *Oleiferous cruciferae.* Annals of Botany, 50(197):85-102.

Anonymous. (2021). National Horticulture Board, Indian Horticulture Database. Ministry of Agriculture & Farmers Welfare, Government of India. www.agricoop.nic.in.

Bailey LH. (1930). Cultivted Brassica II. Gentes Herbarum. 2(5):209-267.

Bhat, K. A., Masoodi, S. D., Bhat, N. A., Ahmad, M., Zargar, M. Y., Mir, S. A. and Bhat, M. A. (2010). Studies on the effect of temperature on the development of soft rot of cabbage (*Brassica oleracea* var. *capitata*) caused by *Erwinia carotovora* sub sp. *carotovora*. *Journal of Phytology*, 2(2):64–67.

Bose, T. K., Kabir, J., Maity, T. K., Parthasarathy, V. A. and Som, M. G. (2002). Vegetable Crops. Bhubani Mitra, Naya Prokash, 206, Bidhan Sarani, Calcutta. 1:345-392.

Chaudhary MM, Bhanvadia AS and Parmar PN. (2015). Effect of integrated nutrient management on growth, yield attributes and yield of cabbage (*Brassica oleracea* var. *capitata* L.) under middle Gujarat conditions. *Trends in Biosciences*. 8(8): 2164-2168.

Debra K R and Misheck D. (2014). Onion (*Allium cepa*) and garlic (*Allium sativum*) as pest controlintercrops in cabbage-based intercrop systems in Zimbabwe. *IOSR Journal of Agriculture and Veterinary Science*. 7(2):13-17.

Goswami L, Nath A, Sutradhar S, Bhattacharya SS, Kalamdhad A, Vellingiri K and Kim KH. (2017). Application of drum compost and vermicompost to improve soil health, growth and yield parameters for tomato and cabbage plants. Journal of Environmental Management. 200:243-252.

Guvenc Iand Yildirim E. (2004). Increasing Productivity with Intercropping Systems in Cabbage Production. *Journal of Sustainable Agriculture.* 28(4): 29-44.

Habiba MU, Rahman MA, Hassan MR. (2009). Effect of sulfur and boron fertilizers on growth and yield of different varieties of cabbage. *International Journal of Sustainable Agricultural Technology*. 5(4): 19-23.

Lee HJ, Lee JH, Lee SG, An S, Lee HS, Choi CK and Kim SK. (2019). Foliar application of biostimulants affects physiological responses and improves heat stress tolerance in Kimchi cabbage. *Horticulture, Environment, and Biotechnology*. 60: 841–851.

Manea AI. (2017). Fertilizer type on cabbage growth and yield. *International Journal of Vegetable Science*. 23(6): 567-574.

Murugiah S, Vidyasagar, Chatterjee SS and Swarup V. (1987). *Indian J. Agric. Sci*. 57:845-847.

Nieuwhoff, M. (1969). Cole Crops-Botany, Cultivation and Utilization. World Crops, Book, Leonard Hill, London.

Olaniyi JO and Ojetayo AE. (2011). Effect of fertilizer types on the growth and yield of two cabbage varieties. *Journal of Animal and Plant Sciences*. 12(2): 1573-1582.

Radovich TJK, Kleinhenz MD and Streeter JG. (2005). Irrigation timing relative to head development influences yield components, sugar levels and glucosinolate concentrations in cabbage. The Journal of the American Society for Horticultural science. 130(6):943–949.

Saha UK, Hasnat MA, Haider J, Saha RR and Kawai S. (1988). *Japanese Journal of Tropical Agriculture*. 42:71-77.

Sarkar A, Rahman S, Roy M, Alam M, Hossain MA and Ahmed T. (2021). Impact of blanching pretreatment on physicochemical properties, and drying characteristics of cabbage (*Brassica oleracea*). *Food Research*. 5(2):393–400.

Sarkar D and Rakshit A. (2021). Bio-priming in combination with mineral fertilizer improves nutritional quality and yield of red cabbage under middle Gangetic plains, India. *Scientia Horticulturae*. 283:1-12.

Singh UP, Syamal NB, Ramkumar. (1978). Effect of different dates of transplanting on growth and yield of cauliflower (*Brassica oleracea* var. *botrytis* L.). *Indian Journal of Horticulture*. 35(2): 138-142.

Swarup, V. (2006). Vegetable Science and Technology in India. Kalyani Publishers. 380-410.

Walker JC and Wellman FL. (1928). A survey of the resistance of subspecies of *Brassica oleracia* to yellows (*Fusarium conglutenance*). *Journal of Agricultural Research*. 37:233-241.

2

Advances in Production Technology of Cauliflower

Suprava Biswal[1]*, Toko Yarin[2] and Dibya Jyoti Mukhia[3]

[1&3] *Ph.D. Research Scholar, Department of Vegetable and Spice Crops, Uttar Banga Krishi Viswavidyalaya, Pundibari, Cooch Behar-736165, West Bengal*

[2] *Ph.D. Research Scholar, Department of Plantation Crops & Processing, Uttar Banga Krishi Viswavidyalaya, Pundibari, Cooch Behar-736165, West Bengal*

**Email: supravabiswalchc2@gmail.com*

Introduction

Figure 2.1: Cauliflower

Cauliflower *Brassica oleracea* var. *botrytis*) is a cold season vegetable cropbelongs to family cruciferous or Brassicaceae family with somatic chromosome number 2n = 18. Cauliflower is derived from the Latin word i.e., 'caulis' means stem and 'floris' means flower and wild progenitor is *Brassica cretica.* In India, cauliflower is grown in two different varieties: snowball type and tropical type. However, at first, other varieties of cauliflower, such as the European and Cornish types from England, were also brought to India.

Later, the Cornish and European forms of cauliflower were intercrossed to develop the tropical type of cauliflower. The Indian type or tropical type of cauliflower can be grown in hot weather and high rainfall condition therefore, such type of cauliflower is mostly grown in Indian plains during the month of May to January followed by temperate type or snowball group based on the temperature based varietal classification.

The eastern Mediterranean region is believed to be the center of origin of the cauliflower crop i.e., mostly in Cyprus or Greece. First time, the cauliflower was introduced in to the India by Dr. Jemson in the year of 1822 from England. The ancient Romans and Greeks were known to have grown cauliflower, and it was later introduced to other parts of Europe by the Moors during the Middle Ages. The first recorded cultivation of cauliflower in North America was in the

1820s, and it is now widely grown all over the world. The domestication of cauliflower likely involved selective breeding of wild varieties over centuries to create the large, white, edible heads that are commonly seen today. It is now widely distributed throughout the world and is a popular food in many cultures. China is the highest producer of cauliflower followed by the India. But it is mostly consumed and popularized by USA country. Annually in India, cauliflower was occupied around 459 thousand hectares area and production of 8800 thousand tones with productivity of 19.17 tons/hectares (Anonymous, 2021-22). In India, cauliflower is mostly produced in West Bengal followed by Bihar, Odisha, Haryana and Chhattisgarh etc.

Economic Importance and Uses

Cauliflower is a rich source of vitamins particularly vitamin C and vitamin K. It also contains antioxidants namely glucosinolates, phenolics compounds, free and total amino acids which may help to protect against different types of cancer and heart diseases also. Additionally, it is full of dietary fiber, which can aid in digestion and weight management. 100 gm of cauliflower contains 29.9% protein, 42.4 mg ascorbic acid, 7.2 mg β-carotene, 66.8 mg iron, 2.6% ionizable iron, 2.3% *in vitro* iron, 1.5 mg copper, 5.6 mg manganese and 4.7mg zinc (Singh *et al.*, 2001). Cauliflower is playing a unique role for preparation of culinary dishes, including soups, stews, and casseroles. It can also be eaten raw, roasted, or sauteed. It can also be used as a low carbohydrate alternative to rice or mashed potatoes, or to make cauliflower "pizza crust" or "tortillas". Cauliflower is also used to make pickles, curry and in various cuisines.

Botany

The plant typically grows to 1-2 feet in height and has large, deeply lobed leaves that form a rosette around the base of the stem. The edible part of the cauliflower is the white, immature flower head, also known as the "curd." The curd is composed of tightly packed clusters of buds or florets that are protected by large, green leaves, called "outer leaves. "The florets of the cauliflower are arranged in a compact, head-like structure called an inflorescence i.e., cymose type inflorescence. The inflorescence is composed of many small flowers, each containing four petals and six stamens which are in tetradynamous condition i.e., 4 long and 2 short stamens. The flowers are typically hermaphrodite, meaning that they can pollinate themselves, but can also be pollinated by insects such as honey bees. Though, flower shows self-incompatibility means both ovule and pollen are viable but due to certain reason of plant and as well as flower factor pollen is unable to fertilize the ovule of the same flower. Self-incompatibility is lowest in early cauliflower cultivar whereas highest in late cauliflower or snowball type cultivar. The curd is typically white, but can also be found in shades of green, purple, and orange. The curd of cauliflower has main stem, primary branches and bracts are there. The curd of the cauliflower has been elaborated by Bailey (1930) as fasciated and hypertrophied flower stalks which produces abortive flower structure. He also concluded that the "grains or papillae" that make up the curd's surface are the flowers, and they are modified, thickened flowers that serve no use. He thought that, like cabbage when the head is gone, the functional flowers appeared on branches from below the curd. Sidki (1962) finally gave a detailed study about cauliflower curd i.e., curdis a 'prefloral fleshy apical meristems'.

Taxonomical Classification

Cole crops has common progenitor and originated from the wild cabbage (*Brassica oleracea* var. *sylvestris* L.). Taxonomy is the science of classification of organisms, which is based on evolutionary relationships. The taxonomical classification of cauliflower suggested by Hooker (1862) was as follows:

Rank	**Scientific Taxonomical Details**
Kingdom	Plantae
Division	Tracheophyta
Class	Magnoliopsida
Order	Brassicales
Tribe	Brassiceae
Sub-tribe	Brasicinae
Family	Brassicaceae

The taxonomy of the genus Brassica and their hybrids has monogenomic species and digenomic species was recorded by Bose *et al.* (2002).

Monogenomic species	**Species**	**Cultivated type**	**No. of chromosome**	**Genomic constitution**
	Brassica nigra Koch.	Black mustard	n=8	bb
	Brassica oleracea L.	Cole group	n=9	cc
	Brassica campestris L.	Turnip group	n=10	aa
Digenomic species	*Brassica carinata* Brown	Abyssinia mustard	n=17	bc
	Brassica juncea (L) Coss.	Brown leaf mustard	n=18	ab
	Brassica napus L.	Swede group	n=19	ac

Climate and Soil

Climate plays a significant effect in the transition from the vegetative to the curding and development stages. Germination is best at temperatures between 10 and 21°C. Since temperature affects growth stages from vegetative to reproductive, it is thermosensitive. The initial growth stages starting from vegetative to curding phase requires temperature between 5°C and almost 28-30°C depending on the variety of a specific maturity group. To transition out of the juvenile period of the plant, a certain quantity of vegetative growth is required. Transition from the vegetative to the curding phase requires 4 weeks for the mid-late group and 5 to 6 weeks for the late maturity group. Initially, young age plant needs 23°C ideal temperature for growth, while 17–20°C is preferable for later stages of growth. The tropical cultivars may survive temperatures as high as 35°C. If the temperature remains greater than what is necessary for curding, the plant will continue to develop vegetatively without producing any curd. While early cultivars will develop curd at the same temperature, late group cultivars need 15-20°C for optimal growth. As a result, tropical cultivars cultivated at lower temperatures become button and exhibit significant riceyness to usual green bud growth. During the curd start phase, the temperature shouldn't vary too much because doing so reduces the quality of the curd. Maximum and lowest temperatures of 15.4-18.9°C and 4.6-6.8°C, respectively, throughout the cold month of the year i.e., Oct-Dec caused good vegetative development in cauliflower in Kullu Valley, according to Gill and Singh (1973). They also discovered that temperature changes in February had a negative impact on seed output.

The ideal temperature range was found to be less than 31.2°C for the maximum temperature, less than 26.8°C for the mean temperature, less than 8.8°C for the diurnal temperature range, less than 6.0 hours of bright sunshine, less than 22.3 MJ m^{-2} for the solar radiation, and less than 22.5°C for the minimum temperature when planting tropical cauliflower. Black rot, a bacterial disease, was discovered to be a severe danger to cauliflower cultivation in this location, and intermittent rainfall and greater relative humidity were shown to be conducive for the prevalence of pests and diseases (Karthika, 2012). In Pedong, Kalimpong, Chatterjee and Som (1990) concluded that 20.1°C to 26.3°C temperature with 90% relative humidity was effective for improvement of curd growth and development. Temperature is an important factor during initial growth period of the curd whereas photoperiod plays an important role during enlargement of curd or maturity period of the cauliflower curd (Wurr and Fellows, 1990).

Cauliflower is a heavy feeder crop requires well-drained, fertile soil rich in organic matter with a pH between 6.0-6.8. The soil should be high in organic matter, as cauliflower requires consistent moisture and good drainage. Adequate irrigation is essential for successful growth, as the plants have a shallow root system and are sensitive to drought. Sandy loam soil with good drainage is the ideal soil for cauliflower cultivation. In acidic soils, magnesium deficiency symptoms can manifest fast, while browning is caused by reduced boron availability at pH levels greater than 7. Early type cultivar of cauliflower in Indian plain mostly in tropical and subtropical area grown in light medium soil condition. But mid-season and late type cultivar can be grown in heavy soil with well-drained soil condition. This crop is not tolerant to soil salinity condition so salinity tolerance level should be lies in between C_2 and C_4 (Bose *et al.*, 2002).

Field preparation

Field preparation for cauliflower planting involves several steps:

- Choose a well-draining, fertile soil with a pH of 6-7 and remove any debris or weeds from the field.
- Bring the soil to a fine tilth and dig holes in hills with a 45 cm separation in either direction. Make 60 cm deep ridges and furrows on plains.
- Incorporate organic matter, such as compost @ 20 ton/ha. of farmyard manure or cow dung into the soil to improve fertility
- Test the soil for nutrient deficiencies and amend as needed.
- Make sure the field has proper drainage and irrigation in place.
- Plant seedlings or seeds at the appropriate time as per the suitable variety at different places in India.

Classification of Varieties

Classification of cauliflower varieties win different groups was suggested by Swarup and Chatterjee (1972) as given under below:

Cauliflower types	**Origin of cultivation**	**Initial period of cultivation**	**Plants morphological character**
Italian	Mediterranean	16th Century	Plants have a short stature, huge erect leaves with rounded bluish green tips, and good unprotected curds.
Cornish	England	Early 19th Century	Plants have strong, long stems and loosely arranged, often wavy leaves, as well as flat, uneven, loose, yellow, very delicious curds.

Roscoff	France	19th Century	Short plants with long, erect, wavy, pointed-tip, bluish-green leaves; white or creamy curds that are hemispherical and well-protected.
Erfurt and snowball	Germany and Netherlands	18th Century	Dwarf plants with upright, short leaves and solid, well-protected curds.
Indian cauliflower	India	Late 19th Century	Plants have short and long stalks, and their leaves are widely dispersed and wavy. The curds are unprotected, flat, yellow to creamy, and flavorful.

Additionally, cauliflower cultivars can be categorized according to maturity. The majority of the early kinds of cauliflower grown in India are referred to as "Kunwari" (mature from Sept to Oct), "Katki" (mature from Oct to Nov), "Agahani" (mature from Nov), "Posi" (mature from Dec), and "Maghi" (mature from Dec) (Jan maturity). For features related to maturity, plant development, and curd, the majority of these cultivars are extremely heterozygous. According to the curd maturity, there are five groups in which the cultivars are divided according the temperature requirement during growing month (Singh and Nath, 2011).

Maturity Time	Variety Name	Temperature Required for Curd Initiation and Development
Early Type		
September	Early kunwari, Pusa Early Synthetic,	20-27ºC
October	Pusa Deepali, Pant Gobhi-2	20-25ºC
Mid-Early Type		
November	Improved Japanese, Pusa Hybrid-2	16-20ºC
Mid-Nov to mid-Dec	Pusa Hybrid-3, Pusa Sharad, Pant Gobhi-4	16-20ºC
Mid-Late		
December	Pusa Synthetic, Pusa Shubra	12-16ºC
Mid-Dec to mid-Jan	Pusa Himjyoti, Punjab Giant-35	12-16ºC
Late Snowball	Pusa Snowball-1, Pusa Snowball K-1	10-16ºC

Singh (2015) observed that early group variety (Pusa Himjyoti) produced a high net curd output when planted at high temperatures, whereas mid-late maturity cultivar (Swati) and late maturing varieties (Pusa Snowball K-1) produced a highest yield when planted at low temperatures. Curd diameter increased with increasing temperature in early cauliflower, however mild temperature range aided in developing big curd size in mid maturity group type. When the temperature was raised from freezing to 10-15°C, the late group type generated the largest curd diameter.

Varieties/Hybrids

Dania

It was developed for the eastern hilly region by IARI, Regional Station, and Kalimpong. Strong plants that produce medium-sized curd. Can be withstand under the abiotic stresses.

Early Kunwari

It is an early variety selected by PAU, Ludhiana, that may be grown in Punjab, Haryana, Himachal Pradesh, and Delhi. The leaves are waxy bloom-heavy with bluish green in color and sowing time for this cultivar is 15th May to end of the May.

Hisar-1

It is a mid-late maturity cultivar with average yield is 240 q/ha.

Improved Japanese

Israel introduced cultivar. Plants are upright with blue-green leaves. Susceptible to high temperature. It will be sown in the plains of northern India between the end of July.

Pant Gobhi-2

It is a composite cultivar. This cultivar often yields 100 q/ha.

Pant Gobhi-3

Synthetic variety generated by crossing of eight katki group inbred lines, and it was distributed by GBPUA&T, Pantnagar. Plants have sparse, semi-erect leaves and relatively longer stems. Curds mature in 110 days with yield of 120 q/ha and non-riceyness type.

Pant Gobhi-4

It is simply recurrently selected by the Aghani group, released by GBPUA&T, Pantnagar. Plants have erect leaves and a medium growth rate. Curds are solid, spherical, and creamy white.

Pant Shubra

Curds are small, slightly conical, cream-colored, non-ricey, and leafless. Rainy season is appropriate for it. Self-blanching type.

Punjab Giant-26

Main or late maturity type cultivar developed by PAU, Ludhiana. Tall plant with a height of about 57cm long. White, solid curd with medium in size.

Punjab Giant-35

It is a late season cultivar with average yield is around 225 q/ha. Curds are snow white in colour, medium size with compact head.

PusaAghani

Nov-Dec is the maturity time of this cultivar. Average yield is 150-160 q/ha.

Pusa Deepali

This is a self-blanching variety is generated through inbreeding breeding technique which is selected from local material at IARI, New Delhi. It is mostly suitable to grow in north Indian condition mainly Punjab and Delhi region. This variety can give optimum yield when it will be sown during end of May to 1st week of June and harvested at month of Oct when temperature is attains at 20-25°C.

Pusa Early Synthetic

This is a resistant cultivar to riceyness. It is mostly cultivated in southern and northern states with yield is 117q/ha.

Pusa Himjyoti

The month of May-Aug are the best timing for sowing in hilly region is the key characteristic of this variety. This is a round type early maturity cultivar takes about 30 DAT to give quality curd with average yield is 160 q/ha.

Pusa Hybrid-2

It takes 120 DAT for maturity of this cultivar. It is a downy mildew resistant cultivar with average yield is 230-250q/ha. This is mostly suitable in a cultivation of Assam-Bengal basins and plains of Ganga alluvial region.

Pusa Katki

It is earliest type cultivar, maturing is in Oct-Nov with sowing time is middle of May.

Pusa Sharad

Plants are having small stalk with semi-erect open types with yield of around 260 q/ha. It takes 85 DAT for harvesting with 750-1000 gm is the standard weight of the curd. It is mostly cultivated in the northern and eastern part of India.

Pusa Shubra

This cultivar is resistant to black rot of fungus and tolerant to riceyness physiological disorder. The average weight of the curd is 700-800 gm with yield of 250-300 q/ha. It takes 125-130 DAT for 50% curd maturation with ideal temperature of 12-16°C.

Pusa Meghna

The average weight of curd is 400 gm. Average yield is 12.5-15 ton/ha.

Pusa Paushija

Curds are compact with average weight is 900gm. Yield of this cultivar is 30-35ton/ha.

Pusa Betakesari

First biofortified indigenously breed beta-carotene rich (800-1000 micro gram/100 gm of curd), orange colored, self-blanching variety with curd weight is 1.25 kg. Average yield is 42-46 ton/ha.

Pusa Snowball-1

This cultivar is developed at crossing between line EC 12013 and EC 12012 released at IARI, Regional Station, Katrain (Kullu Valley) with sowing time is 15th Sept to 31st Oct in northern India. It is a late maturity type cultivar usually require 10-16°C temperature for growth and development.

Pusa Snowball-2

It is a late type cultivar. Harvesting time of this cultivar is end of Jan to early week of Feb at 10-16°C temperature.

Pusa Snowball K-1

This cultivar is tolerant to black rot disease caused by *Xanthomonas campestris* released at IARI, Regional Station, Katrain. Long staying capacity in the field during maturation of curd without hampering its quality is the key characteristic of this cultivar.

Snowball-16

It is a late maturity type cultivar with yield of around 25-30 ton/ha.

Pusa Synthetic

IARI, New Delhi released, mid-season synthetic cultivar generated from seven good combining ability inbred lines. Curd is ready for sowing at 15th Sept to 30th Sept and suitable for harvesting at 15th of Dec to 15th of Jan with optimum temperature of 12-15°C.

Summer King

It is a summer season cultivar usually takes 65 DAT for harvesting of the produce with average curd weight is 400-500 gm.

Kashi Agahani

It is a mid-late maturity group cultivar. Mostly suitable for cultivation in northern India like UP, Punjab and some parts of Chhattisgarh.

Kashi Gobhi-25

This variety is coming under Indian or tropical type group. November is the maturity period where temperature remains around 25°C and suitable for growing in northern part of India.

Arka Spoorthi

This cultivar coming under early type maturity group and developed through using mass-pedigree breeding method. Compact white curd takes 65-75 DAS for maturity with an average yield is 15-16 ton/ha. This variety is resistant to downy mildew and Alternaria leaf spot.

Arka Vimal

Tropical/Indian type cultivar takes around 75-85 DAS for maturity with an average yield 15-18 ton/ha. Resistant to downy mildew and Alternaria leaf spot diseases.

Arka Kanti

This is also an early type Indian cultivar takes around 60-65 DAS for maturation of curd with an average yield of 22-25 ton/ha.

Palam Uphar

Hybrid variety developed at HP state agricultural university, Palampur. Susceptible to downy mildew disease.

Mostly two types of cauliflower are grown in Indian condition that is Indian type or tropical type or heat tolerant type whereas another type is known as

European type or snowball type or Erfurt type or temperate type. There is certain differentiation between Indian type and European type cauliflower is given below:

Indian Type	**European Type**
Tolerant to high temperature	Susceptible to high temperature
Temperature required during curd formation is 20°C	Temperature required during curd formation is varies from 5-20°C
Loose curd with creamish yellow color with strong flavor	Pure white color curd with mild or no flavor denoted as good curd
Long stem with short plant height and freely arranged leaves	Steady plants and leaves are arranged in spirally so that showed as protective jacket to the curd
Heterozygous type (more variable)	Homozygous type (less variable)
Early in maturity	Late in maturity
Small juvenile phase	Long juvenile phase
More self-incompatible	Less self-incompatible
Vernalization is not required in this group but require cold treatment at 10-13°C	Needs vernalization temperature for 8-10 weeks at 7°C

Sowing time, Seed rate and Spacing

There is different sowing time and transplanting time at different maturity crop period. The sowing time of extra early group cultivar is end of Feb with transplanting time at Mar but early type cultivar sowing time is mid of May to end of June with transplanting time at 1st week of July to mid-July. Similarly, mid-early and mid-late type cultivar has sowing time is end of July to end of Aug with transplanting time at 1st week of Sept to end of Sept. At last, late group type cultivar sowing time in the nursery is end of Sept to mid-Oct with transplanting time at end of Oct to mid of Nov. Seed rate is varying according to maturity group of the cauliflower crop likely 600-750 gm seed required for early varieties crop, 500 gm for mid-early season varieties, 400 gm for mid-late varieties and 300 gm for late varieties. After sowing of seed 5 to 6 weeks are usually take for transplanting. During transplanting there is different spacing available according to maturity group namely, 45 cm distance is maintained from row-row and 30 cm distance is maintained for plant-plant in early maturity crop whereas 60 cm X 45 cm is suitable in mid group maturity and late group maturity group of different variety of cauliflower crop. Rahman *et al.* (2007) concluded that among six different plants spacing (30, 35, 40, 45, 50 and 55 cm) of cauliflower, 45 cm plant spacing obtained best growth parameters like highest plant height, curd length, curd diameter with curd yield.

Nursery Management, Method of Sowing and Transplanting

200–500 g of seed would be required to grow one hectare of cauliflower, while the amount needed in the early crops may be closer to1kg.To prepare nursery beds for raising plants, add compost or well-rotted farmyard manure at a rate of 10 kg/m^2. Digging, ploughing, harrowing, and breaking up clods are used to loosen the soil in the beds. About 15 to 20 days prior to seed sowing, sterilizing the soil by soaking nursery beds in formalin at a ratio of 1:49 is helpful in preventing fungal diseases outbreak. After being soaked, the nursery beds should be covered with polythene sheet for about seven days before being dug up again and left exposed for another 5–6 days.so that the seeds are not harmed by the formalin process. For effective drainage, the nursery must be elevated from the ground surface by about 15 cm. Once the sowing of the seed is over the nursery beds are covered with soil and finely sieved any animal-based manure. Before watering, the beds should be properly mulched with something like dry cane grass or polythene. When plants are 15-20 days old, watering can be done by using of water cane over the organic grass mulch and when emergence of seedling occurs mulch can be removed from above the nursery bed. For developing healthy seedlings, nursery beds should frequently be weeded and intercultural. In general, depending on the variety, temperature, and soil richness, seedlings are prepared for transplanting 3 to 6 weeks after seeding. While mid-season and late variety seedlings can be transplanted at three to four weeks old, early crop seedlings that are five to six weeks old have superior establishment and lower damaged of seedlings. Depending on the kind, temperature, and soil, the planting distance may change. Early variety seedlings are often planted 60 cm apart in rows and 30 to 45 cm apart in plants. The spacing for the main season and late variety is 60-75 cm (R-R) X 45-60 cm (P-P). Hossain *et al.* (2020) showed that at plant spacing of 60cm (R-R) X 45 cm (P-P) with 38 days old seedlings transplanted using of Silver Cup-65 cultivar gave highest growth and development character of cauliflower vegetable crop.

Transplanting of seedlings on15th Oct proved beneficial in North East Indian conditions for cauliflower variety Snow Ball-16. However, by transplanting on October 1st and November 1st, yield was marginally reduced (Singh *et al.*, 1978).

Vegetable-based cropping system

Vegetable-based cropping system is an alternative cropping system helps to increase productivity as well as focused on sustainable yield. Yildirim and Guvenc (2005) concluded that increased profit, yield, net income and LER was also more than 1 while cauliflower was taking as a main crop with intercropped were lettuce, radish, snap bean and onion. Sharma *et al.* (2014) recorded that conjoint application of integrated nutrient especially conjoint treatment of manure, fertilizer and PGPR gave highest yield and plant growth attributing traits besides helps in improvement of soil health also in cauliflower-french bean-okra based cropping system. Intercropping of cauliflower with leguminous crop like clover helped in improvement of cauliflower yield in spite the use of reduction amount of nitrogenous fertilizer. Tempesta *et al.* (2019) also observed that the increase in yield of leafy type lettuce as a succeeding crop with requirement of reduction nitrogen fertilizer doses so it might be due to availability of organic matter from the cauliflower-clover based intercropping system.

Nutrition Management

Application of 25-30 ton/ha. of well rotten farmyard manure in the main field at 15-20 days before transplantation of seedlings is recommended for cauliflower crop. In the field, it should be completely blended. Once the soil's fertility state has been established, the appropriate amounts of basic fertilizer doses of NPK should be given. However, according to research done for the All India Coordinated Vegetable Improvement Project, the following recommendations for the amounts of phosphorus, nitrogen, and potassium have been made for different states in India (Anon, 2000).

- Adoption of the variety Snowball-16 under different zone requires different fertilizer doses like 40 kg nitrogen/ha. is recommended at basal and top dressing with 30 kg nitrogen/ha as foliar spray under Sabour climatic condition. However, under Pantnagar conditions, the same variety has been advised to receive 55 kg nitrogen fertilizer doses/ha. whereas 30 kg nitrogen/ha. can be applied as foliar spray.
- In addition to NPK, 15 kg/ha of borax and 15 kg/ha of ammonium molybdate should be applied to soils that are weak in boron and molybdenum, respectively. At the time of transplanting, half dosages of nitrogen and full doses of phosphate and potash are applied. After transplanting, the remaining nitrogen is applied 30–35 days later.

Not only inorganic source of nutrient will be responsible for quality yield of cauliflower crop but also combination of organic and as well as inorganic sources of nutrient can give better curd yield. Pawar and Barkule (2017) concluded that integrated nutrient management gave a better result in terms of curd yield, curd length, curd weight curd diameter etc. in cauliflower cutivar 'Pusa Snowball-16' at the application of 75% RDF + farmyard manure + *Azotobacter* + *Azospirillum*.

Irrigation Management

After transplantation, the first irrigation is administered. Weather, soil type, and variety will all influence how much watering is needed. However, it is crucial to consistently maintain an ideal moisture supply during both the growth and curd development stages. Because of the rainfall, early and midseason crops typically require less watering. Furrow irrigation should be used if planting is done on ridges and there is rainfall. Okasha *et al.* (2022) concluded that maximum cauliflower growth and yield obtained at alternate furrow irrigation system @ 1.43 L/s.

Weed Management

Regular intercultural activities are necessary to aerate the root system and keep the crop free of weeds. Hoeing shouldn't be done too deeply to protect the roots. The penetration of the crust into the root system is typically hampered in medium-heavy and clay soils, which has an impact on plant growth. The roots of plants in ridge planting may become exposed during the wet season, necessitating proper earthing. For the crop to grow healthily, nitrogenous fertilizer should also be used as a side dressing while ridging or earthing. Basalin @ 2-3 kg a.i/ha + 1.2 manual weeding or Tenoran 50 WP @ 1kg a.i / ha can be used for commercial weed control.

Qasem (2007) observed that the maximum cauliflower growth and head yield came with pre-planting oxyfluorfen sprayed at 2.5 L/ha. As compared to other herbicides, the pre-planting applications of pendimethalin (4.6 L/ha), nitrofen (1.4 L/ha), trifluralin (1.5 L/ha), and chlorthal-dimethyl (10 kg/ha) and oxyfluorfen (2 L/ha) boosted the head production of cauliflower.

Integrated weed management is also playing a significant role in cauliflower growth and development. Bana *et al.* (2012) revealed that the plot had black plastic mulch applied showed the best weed control among the various treatments. Although there was no discernible difference between black plastic mulch and rice straw mulch, the growth (plant height, spread, and number of leaves), quality (compactness, colour, and percent protein content of curd), and yield were all higher under black plastic mulch. Among the herbicides pendimethalin in combination with one hand weeding was found most effective.

Blanching

Blanching is a crucial step in the production of high-quality curd in cauliflower because it prevents the curds from becoming yellow from exposure to sunlight. Due to this exposure, the curds may also lose some of their flavor. This issue typically arises in early and midseason maturity group varieties with spreading and open plant types. In certain kinds, the curd is still naturally shielded and encircled by leaf whorls on the inside. This particular process is known as blanching. When curds are fully formed, drawing and tying the tips of the leaves may be used to blanch them. The curd can also be covered with a cauliflower leaf 4 to 5 days before harvest.

Use of plant growth regulator in cauliflower growth and development

Name of plant growth regulators	**Concentrations (mg/l)**	**Method of Application**	**References**
Hydroxybenzoate	10	Seedling treatment	Choonkar and Sharma (1966)
GA_3 + NAA + Boron	100 ppm + 0.2%	Foliar spray at 15 and 30 DAT	Sitapara *et al.* (2011)
GA_3 + Ethrel	150 ppm + 100 ppm	Foliar spray	Khan (2006)

Physiological disorder and their management

There are several physiological disorders with cauliflower, which show themselves as various disease syndromes. While temperature, air, water, humidity, and changes in organic and inorganic diet can also cause physiological diseases, inherited factors are mostly responsible for others. Following is a description of several significant physiological conditions that impact cauliflower:

Riceyness: The elongation of flower buds on peduncles causes curds to form, which are granular, loose, and slightly velvety. Riceyness in cauliflower is a sign of an early floral bud initiation and is regarded as being of inferior grade for marketing. The IARI in New Delhi has discovered that this condition may be caused by any temperature that is either higher or lower than what is ideal for a given cultivar. Mostly riceyness and a delayed harvest early planting of late varieties causes riceyness because of the high temperatures that are common; however, it can also manifest at lower temperatures. Riceyness has also been linked to hereditary factors. Riceyness is also influenced by high relative humidity and a high nitrogen dosage. It can be controlled by use of resistant seed material.

Fuzziness: It manifests as an enlargement of the velvety curds' flower pedicels. The abnormality is both inherited and not inherited. Fuzziness is facilitated by the cultivation of cauliflower outside of its typical season. Fuzziness was reduced by using high-quality seed material during the appropriate season and according to accepted cultural practices.

Leafiness: Common symptoms of this disorders include the curd developing small in size, thin leaves, which lowers the curd's quality. Inheritable or nonheritable variables may cause extremely tiny green leaves to grow in between the curd segment. High temperatures frequently present during the curding phase exacerbate leafiness. More than others, certain types are sensitive to leafiness or bracketing. By choosing types according on how well they adapt, it can be managed.

Brown rot: Boron deficiency is the main cause of this disorder due to acidic pH. It can be identified by the immature, fragile, dark green leaves that develop on it. The old leaves often drop off and are puckered and chlorotic. When there is a severe boron shortage, elder leaves will occasionally curl downward and blister. The developing point could perish, and the leaves continue to be little. At later stage, light brown to dark brown color water-soaked patches developed on the stem and branches of plant results formation of cavities and a hollow stem. which eventually turn rusty brown in colour. The curds that are damaged are tiny and develop a bitter flavor. Applying borax, sodium borate, or sodium tetra borate to the soil @ 20 kg/ha can control this. Spray a 0.25 to 0.50% solution of borax at a rate of 1 to 2 kg/ha in cases of acute deficiency, depending on growth, soil reactivity, and the severity of the deficiency.

Whiptail: It is caused by molybdenum deficiency in very acidic soils. Because the uptake of molybdenum, which is uncommon when the soil pH is 5.5 or higher, is hindered by high manganese concentrations in such soils. The young cauliflower plants start to turn white and chlorotic, especially where the leaf margins meet. Additionally, they cup and wilt. The leaves are ruffled and deformed, and the leaf blades fail to develop normally. Older plants usually only have one or two big bare midribs on the lamina of newly developed leaves, giving rise to the colloquial name "whiptail." By adding lime or dolomite limestone, you can raise the pH of the soil to 6.5 or higher. A soil application of sodium or ammonium molybdate at a rate of 1-2 kg/ha can also suppress cauliflower's "whiptail."

Buttoning: Buttoning in cauliflower refers to the development of little curds with insufficient foliage. Premature heading is another name for it. The leaves are so tiny that they cannot fully enclose the developed head. The transplanting of seedlings older than six weeks old is one of the causes of buttoning. Due to their poorly established root systems, over-aged seedlings transplanted into the field typically take longer to establish. Because seedlings are crowded together in nursery beds for prolonged periods of time, a poor root system growth is conceivable due to insufficient nutrient availability and insufficient room for root system development. Such old seedlings typically have poor development rates.

i) When temperatures are below ideal during the growing season, maturity is delayed.

ii) Planting a late variety as an early variety result in buttoning.

iii) Hot, dry climate is one of the major factors for buttoning results development of small button like curd.

iv) Another limiting factor is availability of optimum moisture content which leads to development of early curd without attaining proper size and weight.

v) Transplant of seedlings at right stage with right time in the main field is a major factor for avoidance of buttoning disorder because delay sowing of seedlings in the main field is crucial for the development and expansion of typical curds.

vi) Slow plant growth in the nursery, excessive crowding, insufficient water, lack of weeding, poor soil quality, excessive salt concentrations, and low-lying areas or fields with shallow and weak top salt can all contribute to buttoning.

vii) Nursery plants that have been vigorously grown and have thicker stems and sessile foliage that are already generative tend to generate buttons in the field.

The slowdown in growth could be brought on by diseases, insects, or root damage (especially *Rhizoctonia spp.)*.

a) It can be regulated by taking good care of the nursery to prevent any obstructions to plant growth.

b) Applying enough nitrogen and water is necessary.

c) Avoid water logging and overpopulation

d) Avoid delaying of transplanting of seedlings

e) All the inter cultural practices should be carried out throughout the crop growth period.

Blindness: It refers to a plant that has no terminal buds or one that has early growth point collapse, no terminal buds, and blindness. Any action that interferes with the formation of the terminal bud might cause blindness. It affects overwintering plants. Plant develops but does not form or curd a terminal bud. Due to the buildup of carbohydrates, it is distinguished by the enormous, thick, dark green, leathery leaves that emerge. Axillary buds occasionally grow, but the plant is unable to produce a curd that is suitable for market. The primary causes of blindness include low temperatures when plants are young, damage to the terminal bud during handling, or injury from insect pests. By taking care of seedlings and avoiding exposing young plants to low temperatures, it can be controlled. Seedlings are protected from insect and pest damage when handling and planting.

Chlorosis: When cauliflower is cultivated on extremely acidic soils, magnesium shortage results in chlorosis. The lower leaves have yellow mottling and interveinal chlorosis. The damaged leaves become rigid and bronze in color. Abscission of the lower leaves, which produces little curds, happens in severely deprived plants. Applying 300 kg/ha of magnesium oxide or liming the soil with dolomite limestone to raise the pH to 6.5 are two effective control methods. Utilizing a fertilizer with soluble magnesium helps to control it.

Hollow stem: It might be caused by an excess of nitrogen nutrition and a lack in boron. When nitrogen causes hollowness, the stem is perfectly transparent and shows no signs of disintegration in contrast to boron's hollowness, which can be distinguished by water-soaked and discolored tissue. Spraying borax at a rate of 0.1 to 0.3 or applying 15-20 kg/ha of borax to the soil can both be used to control it. Boron shortage may use hollow stems. The plants can be spaced closer together or the fertilizer doses can be decreased, nevertheless, for the typical sort of hollowness.

Frost Injury: Young seedling leaves of cauliflower turn both surfaces yellowish-white. Petioles turn white and flaccid, and there may be damage to the stem, parenchyma nearby, and midrib as well. Cauliflower curds that are fully formed are more susceptible to damage from frost than smaller ones. Similar symptoms also occur in Brussels sprouts and cabbage also. It can avoid by spraying of water or by producing smoke in the field to maintain the temperature little bit higher during frost bite condition of the standing crop.

Major diseases and their management

Stalk rot (*Sclerotia sclerotiorum*)

When infected, leaves shed early, turn light green, and eventually change to yellow. On petioles and mid-rib contacting, black soft rot frequently develops later. A fluffy fungus growth has taken over the soil. Stem at ground level exhibits girdling. Ideal circumstances for the onset and spread of disease are cool (20–25°C) and 95%–100% humidity.

Control measures- Some cultural practices can also be used for reduction of diseases that is crop rotation or spraying of chiptox @ 0.02% have also found to be effective against diseases.

Downy mildew (*Perenospora parasitica*)

Symptoms- Plants can become infected at any stage during their growth cycle. Young plants are highly susceptible show white mildew growth on the underside of the leaf. Later on, white mildew growth enlarged and become yellowish color on the upper side of the leaves. A juvenile leaf may fall off.

Older leaves normally survive, and infected regions grow in size as they turn scorching in appearance. When the disease becomes severe, the entire leaf dies and affects the photosynthesis system in the plant and ultimately plant dies.

Control measures- The effective control measure of this downy mildew growth can be spray with copper oxychloride @ 0.3% at the seedlings stage in the nursery bed.

Black Spot (*Alternaria brassicae*)

Symptoms- The initial sign is a little dark spot on the surface of the leaf. Concentric rings form as the spot expands. The infected spot grows in size and defoliates plants. It is a seed crop disease that destroys pods and seeds.

Control measures- This type of fungal infection can be prevented by foliar spray with the captan, thiram @ 0.2% or copper oxychloride @ 0.5%.

Black leg (*Phoma lingam*)

Symptoms- It occurs in the majority of regions, particularly in areas with high rainfall throughout the growing season. Because the fungus is transported by the seed, it might appear at any stage. When the damaged plant's stem is severed vertically, the sap stream is severely discoloured. The entire root system decays from the ground up. The afflicted plants frequently collapse in the field.

Control measures- The main source transfer of diseases through seeds can be avoided by hot water treatment of the seeds. Spraying with copper oxychloride or an organomercuric chemical can be help in prevention of this disease and using resistant varieties will be effectible for removal of disease.

Black rot of crucifers (*Xanthomonas campestris*pv. *campestris*)

Symptoms- Infected plants grow stunted, with one side of the cotyledon turning yellow to black and bending down and dropping off early. Many lower leaves fall off early as well. The surviving leaves may yellow and have dark veins. From the leaf margin, 'V' shaped chlorotic to yellow lesions develop. The veins and veinlets become dark, and the leaf tissue becomes necrotic and brittle. The lesion progresses towards the midrib, and the infection spreads farther into the stem and root as a result of this systemic infection. Vascular tissues in affected areas turn dark, followed by internal breakdown of squishy tissue. Succulent seed stocks and siliqua may acquire black sunken patches.

Control measures- This type of malady can be prevented by spraying of plantomycin @ 100 ppm or Agrimycin @ 100 ppm.

Club root (*Plasmodiophora brassicae*)

Symptoms- Club root infection is mostly prevalent in cole crops. It is mostly seen in saline affected soil where pH is higher than 7. Affected root shows irregular galls in the tap root system whereas spindle shape in the lateral roots ultimately it affects the plant growth results wilting of the plant occurs.

Control measures-Treated with 4% calomel paste of the seedlings root at the time of transplanting time can be effective for the control of this diseases.

Major pests and their management

Leaf webber (*Crocidolomia binotalis*)- The eggs present in a mass and joint together by production of viscous glue type substances of this pest mostly seen under the leaf surface. Adult moths are mostly slender type and light brown in color.

Control measures- Key natural mortality causes for leaf webber were observed in late larval instars as parasitization by *Braconhebetor*, i.e., *Palexorusta solennis* (Singh *et al*, 1978).

Mustard sawfly (*Althalia lugens proxima*)- In the early stages, the blackish caterpillar targets all cole crops and feeds on the leaves of immature seedlings. When contaminated leaves are touched, they curl and fall to the ground.

Control measures- Hand picking caterpillars is the greatest way of controlling the larvae if the infestation is modest and the number of attacked plants is limited.

Mustard aphids (*Lipaphis erysimi*)- When the weather is cloudy, this little greenish white bug attacks cabbage and other cole crops. They feed on the sap of the plants. Plants wither and die as the damaged leaves curl.

Control measures- Aphids is the serious pest for plant due to infection starts from the nursery bed and can be controlled effectively by spraying of rogor, metasystox etc. @ 1ml/lt. of water starting from the nursery stage at 15 days interval. The other method can be used to control the aphid infestation that is soil incorporation of thimmet @ 12-15 kg/ha. in two separate doses like half portion of the recommended doses at planting time and rest half portion at the time of earthing up.

Diamond back moth (*Plutella xylostella* L.)- It was originally seen on cruciferous vegetables in India in 1914. A little pale green caterpillar that eats on leaves and chews on them. The contaminated produce is no longer marketable.

Control measures- This pest infestation can be reduced by use of trap crop that is mustard crop and that should be planted 15 days earlier before main crop that is cabbage crop. Foliar spray twice or three times with cypermethrin (0.2 kg a.i. /ha), femvalerate (0.1 kg a.i./ha), or sulprofos (1.0 kg a.i./ha). It can also be managed by foliar spraying of padan @ 500gm/ha. in cabbage at 15 days interval starting from the leaf primordial stage. Other type of pesticides can also be effective to control of DBM pest by spraying of cypermethrin @ 0.2 kg/ha. or femvalerate @ 0.1 kg/ha.

Harvesting, Yield and Storage

The curds are harvested as soon as they reach the proper age, are compact, and preserve their original white color. Delaying the harvest causes the curds to overmature, which lowers their quality. These curds could become sloppy, leafy, ricey, or fuzzy. While preparing the produce for market, the overripe curds should be removed. With a sharp cutting knife or sickle, the curds are sliced off the stalk well below the curd for harvesting. Depending on the variety, maturation group, and season, the yield of the cauliflower varies substantially. The typical yield of early maturing cultivars ranges from 80 to 120 /ha. Cauliflower grown throughout the main season yields 150–200 q/ha. The largest yield, 250–300 q/ha, was produced by the Snowball group and other late maturing cultivars because of their very compact curds and higher plant density per square meter. For 2-4 weeks, cauliflower can be properly stored at 0°C and 90–95% RH.

Future Prospect

Cauliflower is a popular vegetable that has been consumed for centuries, and with the advancement of technology, there are many potential areas for improvement in its production. Here are some possible future prospects for advanced production technology of cauliflower:

- Precision agriculture: Precision agriculture involves using sensors and other technologies to collect data on plant health, soil moisture, and other factors to optimize crop growth and yield. With precision agriculture, farmers can monitor cauliflower crops in real-time and adjust irrigation and fertilization levels to ensure optimal growth.
- Plant breeding: Plant breeding can be used to develop cauliflower varieties with higher yields, better disease resistance, and improved quality traits such as flavor and texture. Advances in genetic engineering and gene editing may enable more rapid development of new cauliflower varieties with desirable traits.
- Controlled environment agriculture (CEA): CEA involves growing crops in a controlled environment such as a greenhouse or vertical farm. With CEA, farmers can optimize growing conditions such as temperature,

humidity, and lighting to maximize yield and quality. This technology can be particularly useful for growing cauliflower in regions with harsh weather conditions.

- Robotics and automation: Robotics and automation can be used to automate various aspects of cauliflower production, such as planting, harvesting, and sorting. This technology can improve efficiency and reduce labor costs for farmers.
- Sustainable agriculture: With growing concerns about the environmental impact of agriculture, there is increasing interest in sustainable production methods for cauliflower. This may include using organic farming methods, reducing water and fertilizer use, and minimizing pesticide use.

Overall, these advanced production technologies can help improve the efficiency and sustainability of cauliflower production, increase yields, and produce higher quality crops.

References

Anonymous. (2000). National Horticulture Board, Indian Horticulture Database. Ministry of Agriculture & Farmers Welfare, Government of India. www.agricoop.nic.in.

Bailey LH. (1930). Cultivted Brassica II. Gentes Herbarum. 2(5):210-267.

Bana, M. L., Kaushik, R. A., Bana, R. S.and Dhakar, M. K. (2012). Integrated weed management in cauliflower cv. 'Pusa Snowball K-1'. *Ann. Agric. Res*, *33*(3), 163-169.

Bose, T. K., Kabir, J., Maity, T. K., Parthasarathy, V. A. and Som, M. G. (2002). Vegetable Crops. Bhubani Mitra, Naya Prokash, 206, Bidhan Sarani, Calcutta. 1:345-392.

Chatterjee, R., & Som, M. G. (1990). Effect of sowing date on curd production and seed yield of cauliflower. *Vegetable Science*, 17(1), 66-69.

Chhonkar VS and Sharma SB. (1966). Studies on the effect of starters and plant growth regulator on growth, yield and quality of cauliflower (*Brassica oleracea* var. *botrytis*L.). *Indian Journal of Horticulture*. 23(1&2):81-87.

Gill HS and Singh JP. (1973). Effect of environmental factors on seed production of late cauliflower in Kullu valley. *Indian J. Agri. Sci.* 43(3): 234-236.

Hooker, J. D. (1862). Genera Plantarum. Eds. G. Bentham and J D Hooker, 1: 57-102.

Hossaina B, Ruhib RA and Mohsin GM. (2020). Effects of varieties and seedlings age on growth and yield of cauliflower. *Tropical Agroecosystems*. 1(2): 75-79.

Karthika, V. P. (2012). *Crop weather relationship in cauliflower (Brassica oleracea var. botrytis L.)* (Doctoral dissertation, Department of Agricultural Meteorology, College of Horticulture, Vellanikkar, Kerla Agricultural University.

Zia, M. I. K. (2006). *Effect of Plant Growth Regulators on Growth Curd Yield and Seed Production of Cauliflower* (Doctoral dissertation, Department Of Horticulture And Postharvest Technology Sher-E-Bangla Agricultural University Sher-E-Bangla Nagar, Dhaka -1207.

Okasha, A. M., Deraz, N., Elmetwalli, A. H., Elsayed, S., Falah, M. W., Farooque, A. A. and Yaseen, Z. M. (2022). Effects of Irrigation Method and Water Flow Rate on Irrigation Performance, Soil Salinity, Yield, and Water Productivity of Cauliflower. *Journals of Agriculture*, *12*(8):254-262

Pawar R and Barkule S. (2017). Study on effect of integrated nutrient management on growth and yield of cauliflower (*Brassica oleracea* var. *botrytis* L.). *Journal of Applied and Natural Science*. 9 (1): 520 -525.

Qasem JR. (2007). Weed control in cauliflower (*Brassica oleracea* var. *botrytis* L.) with herbicides. *Crop Protection*. 26(7): 1013-1020.

Mujeeb-ur-Rahman, M. I., Jilani, M. S., & Waseem, K. (2007). Effect of different plant spacing on the production of cauliflower (Brassica oleraceae var. Botrytis) under the agro-climatic conditions of DI Khan. *Pakistan Journal of Biological Sciences*, *10*(24), 4531-4534.

Rahman M, Iqbal M, Jilani MS and Waseem K. (2007). Effect of different plant spacing on the production of cauliflower (*Brassica oleraceae* var. *botrytis*) under the agro-climatic conditions of D.I. Khan. *Pakistan Journal of Biological Sciences*. 10(24): 4531-4534.

Sharma, A., Sharma, R. P., Sharma, G. D., Sankhyan, N. K., & Sharma, M. (2014). Integrated nutrient supply system for cauliflower-French bean-okra cropping sequence in humid temperate zone of north-western Himalayas. *Indian Journal of Horticulture*, *71*(2), 211-216.

Sidki, S. (1962). Variability studies in cauliflower for economic character. *American J. Bot*, *49*, 290-297.

Singh, D. N. and Nath, V. (2011). Varieties and Hybrids of Vegetables. Satish Serial Publishing House, Delhi. 1:426.

Singh, G., Kawatra, A., and Sehgal, S. (2001). Nutritional composition of selected green leafy vegetables, herbs and carrots. *Plant Foods for Human Nutrition*, *56*, 359-364.

Singh G, Kawatra A and Sehgal S. (2001). Nutritional composition of selected green leafy vegetables, herbs and carrots. *Plant Foods for Human Nutrition*. 56: 359–364.

Singh TB. (2015). Studies on effect of temperature on curd yield under year-round production system of cauliflower (*Brassica oleracea* var. *botrytis* L.) under mid hills of Himachal Pradesh. *International Journal of Bio-resource and Stress Management*. 6(1):140-144.

Singh UP, Syamal NB, Ramkumar. (1978). Effect of different dates of transplanting on growth and yield of cauliflower (*Brassica oleracea* var. *botrytis* L.). *Indian Journal of Horticulture*. 35(2): 138-142.

Sitapara HH, Vihol NJ, Patel MJ, Patel JS. (2011). Effect of growth regulators and micro nutrient on growth and yield of cauliflower cv. 'Snowball-16'. *Asian Journal of Horticulture*. 6(2):348-351.

Swarup V and Chatterjee SS. (1972). Origin and genetic improvement of Indian cauliflower. *Economic Botany*. 26(4): 381-393.

Wurr DCE and Fellows JR. (1990). The influence of field environmental conditions on the growth and development of four cauliflower cultivars. *J. Hort. Sci*. 65(5): 565-572.

Yildirim, E., and Guvenc, I. (2005). Intercropping based on cauliflower: more productive, profitable and highly sustainable. *European Journal of Agronomy*, *22*(1):11-18.

Tempesta, M., Gianquinto, G., Hauser, M., and Tagliavini, M. (2019). Optimization of nitrogen nutrition of cauliflower intercropped with clover and in rotation with lettuce. *Scientia Horticulturae*, *(246)*:734-740.

3

Advances in Production Technology of Knol-Khol

Suprava Biswal[1]* and Toko Yarin[2]

[1] Ph.D. Research Scholar, Department of Vegetable and Spice Crops, Uttar Banga Krishi Viswavidyalaya, Pundibari, Cooch Behar-736165, West Bengal

[2] Ph.D. Research Scholar, Department of Plantation Crops and Processing, Uttar Banga Krishi Viswavidyalaya, Pundibari, Cooch Behar-736165, West Bengal

**Email: supravabiswalchc2@gmail.com*

Introduction

In India, knol-khol (*Brassica oleracea* var. *gongylodes*) is belongs to family brassicaceae with chromosome number is 2n=18. It is grown in north eastern states of India, J &K, West Bengal, Assam, Punjab, and other regions of southern India also. It is distinguished by the formation of knobs(tuber) is a thickening of the stem tissue above the cotyledons. This knob is picked for human use as a raw or cooked vegetable, while young leaves are also utilized in some areas.

Figure 3.1: Knol-Khol

Economic Importance & Uses

Knol-khol, also known as kohlrabi or German turnip, is a biennial vegetable that belongs to the same family as cabbage and broccoli. It is a popular crop in many parts of the world, including Europe, Asia, and the Americas. Knol-khol is a nutritious vegetable that is rich in vitamins C and K, as well as potassium and fiber. It can be eaten raw or cooked, and is often used in soups, stews, salads, and stir-fries. It is also used in the culinary industry as a versatile ingredient in various dishes. It can be pickled, sautéed, roasted, or pureed to make sauces, dips, and spreads.It can be used as a feed for livestock and other animals. It is high in fiber, which makes it a good source of roughage for ruminants such as cows and sheep. Knol-khol can be used as a source of biofuel. It has a high sugar content, which can be fermented to produce ethanol. It is also grown as an ornamental plant, prized for its unusual shape and attractive foliage. It is often used as a border plant or as a specimen in mixed garden beds.

Botany

Knol-khol, also known as kohlrabi, is a vegetable that belongs to the Brassicaceae family, which includes other vegetables such as broccoli, cauliflower, and cabbage. It is a cool-season crop and is commonly grown in temperate regions. The plant has a stout stem that grows up to 60 cm in height and has large leaves that are alternate, lobed, and slightly hairy. The stem of the plant is edible and has a bulbous shape, which can range in color from light green to purple. It is a biennial plant, which means it completes its life cycle in two years. In the first year, the plant grows leaves, stems, and roots. In the second year, it flowers, produces seeds, and dies. However, in most cases, knol-khol is grown as an annual plant, which means it is harvested in the first year before it has a chance to flower and produce seeds. The flowers of knol-khol are small, yellow, and grow in clusters at the tip of the stem. The plant is self-fertile, which means it can pollinate itself and produce seeds without the need for cross-pollination from another plant. The seeds of knol-khol are small, brown, and have a pungent taste. For improving the pollination work some farmers used some attractant for pollinators. Amin *et al* (2016) used pollinator attractants geraniol @1%, citral @1%, geraniol 1%+citral 1%, fruit boost @1% and 0.5% and sugar solution @10% were evaluated for their effect on yield parameters of seed on knolkhol. Application of pollinator attractants enhanced seed yield significantly and highest yield was recorded from plot sprayed with fruit boost @1% with 104.79% increase over control plot. The germination was highest for the same treatment with 10.47% increase over the control.

Taxonomy

Knol-khol (*Brassica oleracea* var. *gongylodes*) is a vegetable crop that belongs to the Brassicaceae family. It is also known as Kohlrabi or German turnip. The taxonomy of knol-khol is as follows:

Rank	**Scientific Taxonomical Details**
Kingdom	Plantae
Subkingdom	Tracheobionta
Superdivision	Spermatophyta
Division	Magnoliophyta
Class	Magnoliopsida

Subclass	Dilleniidae
Order	Capparales
Family	Brassicaceae
Genus	*Brassica*
Species	*Oleracea*
Variety	Gongylodes

Climate and soil

Knol-khol is a cool type crops usually grown in humid climate. In temperate countries, early bolting usually occurs in early type cultivars whereas in sub-tropical condition, fluctuation of day and night temperature is rarely falling below 10°C to carry out vernalization process. Due to high temperature, seeds cannot be set properly after vernalization. Knol-khol can be grown properly at an average monthly temperature of 15-20°C. Low temperatures do not stimulate bolting in late varieties in the early stages. When the plant has grown adequately, it is vernalized in the same manner as other biennial cole crops. Knol-khol is thrives better in frost condition as compared to other cool season vegetable crop. In chilling place some bolting resistant cultivars do not produce knob due to absence of vegetative stage. The round and flat round variants develop long oval-shaped knobs when the temperature is low during the knob creation stage. Low temperatures or frost conditions can cause the development of anthocyanin colors on knobs or plant components. If there is a deficiency of nitrogen and phosphorous in the soil, these pigments become more intense, lowering the quality of knobs, particularly green varieties. Knol-khol can be grown in any soil type. Excellent knobs are produced by soil that is rich in organic matter. Sandy loam type soil is suitable for early crops, whereas clay type or silty loam type soil texture is good for late duration maturity crop due to holding of organic matter is high in heavy soil. The ideal soil pH range for knol-khol cultivation is 5.5-6.8. It also requires full sunlight and cool temperatures for optimal growth. The plant is relatively easy to grow and is not susceptible to many pests or diseases.

Varieties

Golith White (Sadashiv), Sutton, Early White Vienna, Early Purple Vienna, Verma, Pocha and King of the Marketare some of the promising types available in India. But mostly cultivated variety in India are following:

1. **King of North-**Flattish-round shape knob takes an around 60-65 DAT for maturity. The horizontal posture of the lower leaves distinguishes these early types. Late variants reach maturity 75-105 days after planting. Purple Speck is the earliest variation in this group, while Goliath is the latest.

2. **Large Green**- White flesh knob takes 76 DAT for maturity with an average yield potential is 25-28 ton/ha. It is mostly suitable grown in mid and upper hills of western Himalayas region.
3. **Purple Vienna-** Purple color knob having white color flesh takes about 55-65 DAT for maturity. It is usually harvested after 1 week of white Vienna maturation and gives more yield also as compared to White Vienna.
4. **White Vienna-** Medium size knob takes around 55-65 DAT for maturation. Its plants are small, with medium green leaves and stalks. It has a potential yield of 17-23 ton/ha. Dwarf plants with short crowns and globular spherical knobs characterize Early White Vienna.
5. **Early White Vienna-** Dwarf, early type knob round in shape having medium green leaves. It takes 50-55 DAT for maturation.
6. **Early Purple Vienna-** Early type cultivar having purple color leaves and skin of the knob but flesh is green in color. It takes 50-60 DAT for harvesting.

Seed sowing and Nursery bed preparation

All the cole crops are propagated by seed including knol-khol crop with seed rate of 1-1.5 kg/ha. Before sowing in disease-prone areas, seed should be treated with hot water (50°C) for 30 minutes to avoid black rot and Apron 35 @ 2g/kg of seed against downy mildew diseases. In the nursery beds, seedlings are raised. Seedlings that are 4-6 weeks old are ready for transplantation. Nursery beds are typically 60 cm broad and 2.5m long. For a 1m^2 nursery, 45g of fertilizer combination containing 15g of nitrogen, 15gm of phosphorous, and 15 gm of potassium and 2.5-4 kg of farmyard manure mixed thoroughly in soil and a 30cm channel along with the nursery must be prepared. Sunken nursery beds are preferred in light and drought-prone soils. Lime should be applied to acidic soils. The nursery beds should be formalin-treated to reduce seedling damage (40% formaldehyde mixed in 5-6 parts water). This solution is saturated into the soil, requiring 5 liters/m^2. Formaldehyde treated nursery beds should be covered with polythene, burlap or canvas for 2days for proper reaction in the sealed disease infested nursery bed and soil should be aerated by removing polythene cover for 4 days of before sowing of seed. This treatment can be substituted by soaking the soil with captan (0.3%). Seeds are placed in rows 5-6 cm apart to facilitate hand hoeing, weeding, and thinning. When seedlings are planted too close together, they are susceptible to damping off disease and become slender. Proper spacing produces stout and vigorous seedlings. A depth of 1.5-2cm is ideal because deeper seeding inhibits germination. The nursery bed is grass-covered to conserve moisture and ensure uniform germination. It is irrigated as needed with a watering can. To control damping off, the mulch is removed soon before seed germination, and a drenching with Dithane M-45 (0.2%) is recommended. When seedling development is poor, urea (nitrogenous fertilizer) may be added to the spray. Excess N, on the other hand, results in sensitive, tall plants with poor establishment following transplantation. To avoid transplanting shock of the newly emerged seedlings can be hardened in the nursery for about a week before the main transplanting date on the main field.

Transplanting

Planting can be done in Sept on the plains of north India, while it is best done in Oct in the cooler winter regions. Seeds are sown in the hills of northern India from Mar to Aug. 5-6 weeks old seedlings are suitable for transplantation for summer and autumn crops. Low-cost polyhouses can be prepared for nursery management during Mar and Apr for protection against cold and frost. To achieve the fine tilth, 2-3 ploughings with one harrowing by using disc harrow is necessary. The beds and channels have been prepared for irrigation. Seedlings are transplanted in the evenings and/or on cloudy days. This should be followed by certain light irrigation. 5-6 days after transplanting, damaged plants should be removed and gaps closed. Depending on the climatic conditions and soil fertility, transplanting is done at a closer spacing of 25 cm - 30 cm (R-R) X 25 cm – 45 cm (P-P). Close spacing increases yield while decreasing knob size. Early types can be planted closer together, whereas late varieties require broader spacing. Silatar *et al.* (2018) concluded that plant spacing of 20 x 20 cm^2 with variety of Purple Vienna gave good knob growth but for quality 30 x 30 cm^2spacing showed better as compared to other plant spacing under north Gujarat climatic conditions.

Manuring and fertilizers

Knol-khol is a heavy feeder and responds well to manuring. A yield of 200 quintals/ha. removes around 100kg of nitrogen, 85 kg of phosphorous, and 170kg of potassium. Excess application of nitrogen fertilizer may result in excessive foliage growth and a delayed yield. It is preferable to apply N in two stages. At the time of transplanting, half of the N as well as the full amount of P and K are applied. The remaining part of two equal split of nitrogen fertilizer doses applied at 2 different crop growth period 1st at 3 weeks after transplanting of seedlings and rest part should be given at knob development stage. Farmyard manure is mixed into the soil 1 month before transplanting. Maximum N and K levels are required for flavorful knobs. As like macro nutrients, micro nutrients are also required for complete the life cycle of knol-khol. Some of the micro nutrients like boron and molybdenum which is essential for cole crops and helps in avoidance of physiological disorders like browning and whiptail. Foliar application of Multiplex (0.2-0.3%) might be helpful to correct the micronutrient deficits. However, boron can be applied in the field @ 10-15 kg/ha as a borax form or foliar spraying of borax @ 0.3% is also effective to the control of browning malady. Whiptail physiological disorder is caused by molybdenum deficiency which mostly prevalent in acidic soil can be overcome by reclamation of soil with lime or soil treatment of ammonium molybdate @ 200-300 g/ha at a concentration of 0.01-0.1%.

Mishra *et al.* (2014) conducted an experiment using of organic and inorganic nutrient combination through applying integrated nutrient management system. They concluded that treatment combinations of recommended doses of NPK (150:38:63 kg/ha.), vermicompost @ 2.5 ton/ha. and biofertilizer (*Azotobacter*, *Azospirillum* and phosphorous solubilizing bacteria) application gave good growth with higher quality of knol-khol crops. Kumar *et al.* (2018) studied on effect of organic, inorganic, biofertilizers (*Azospirillum*) and their combinations on White Vienna variety of knol-khol crops and they observed combination of biofertilizers with RDF of inorganic nutrient gave good growth but improved quality recorded at application of neem cake with inorganic fertilizer.

Aftercare

It is critical to maintain steady growth. Any delay in growth results in fibrous and woody knobs. On the other hand, despite there is a lot of varietal variation, too quick growth following slow initial growth may result in cracked knobs. Due to a lack of light, the knobs in close spacing may become extended, as in knobs growing in the shade. Sometimes excess application of nitrogen doses with high temperature causes for elongated knob. Long dry spell with immediate water supply causes cracked knob which is not utilized for selling in the market.

Weeding

Intercultural operations are carried out primarily to control weed development, loosen the soil, and maintain adequate moisture levels. Hoeing is a main intercultural operation helps to avoid the crop-weed competition by uprooting the weed population growth in the main field. Crop-weed competition at any crop growth stages affects the crop growth results poor yield. Regular hoeing help keep the weed population under control. Shallow hoeing should begin as soon as the weeds appear. Once the weeds have established themselves, removing them affects the root system of the plants, resulting in poor development.

Hoeing is interrupted as soon as the soil is covered with foliage. Hoeing is done by using khurpi garden tool should be avoided at the knob development stage so that weeding can be done by using human labor is necessary. All the monocot and dicot weeds population before transplanting in the main field can be reduced by the spraying of treflan (0.5 lt./ha trifluralin). It is also possible to use black polythene mulch to control weeds. The effectiveness of Stomp (Pendimethalin @ 1- 2 kg/ha) in weed control has recently been recommended. Weedicide application coupled with 1 or 2 hand hoeing is a beneficial and cost-effective practice.

Irrigation requirement

To get uniform and healthy knob optimum application water is required. According to soil and weather conditions, irrigation can be done in the field. Irrigation at 15 day intervals is sufficient. Irrigation should be kept to a minimum. Water can be applied in the field when soil moisture content falls 50% below of the field capacity. Irrigation is harmful during the maturation of knobs.

Harvesting

Before they have fully matured, the knobs are harvested by cutting the stem just below it with a sharp knife or sickle. They then become rough and woody. Knol-khol should be harvested when it reaches maturity, which is usually around 60-80 days after planting. The crop should be harvested when the stem is about 8-10 cm in diameter. It is important to handle the crop carefully to avoid any damage or bruising. The root portion of the plant is removed and the plants, together with the tender leaves, are bundled in bunches for market. It is also marketed once the leaves and roots have been removed.

Post-harvest Management

Knol-khol, also known as kohlrabi, is a cool-season vegetable crop that requires proper post-harvest management to maintain its quality and prolong its shelf life. Here are some post-harvest management practices that can help preserve the quality of knol-khol:

1. Cleaning: After harvesting, the knol-khol should be cleaned to remove any soil or debris. This can be done by washing the crop in clean water or wiping it with a damp cloth.
2. Sorting: The knol-khol should be sorted to remove any damaged or diseased parts. Any knol-khol with cracks, cuts, or bruises should be removed as they are more susceptible to spoilage.

3. Cooling: Knol-khol should be cooled immediately after harvesting to prevent any further growth and to maintain its freshness. The crop can be cooled by placing it in a cool, well-ventilated area.
4. Storage: Knol-khol should be stored in a cool, dry place with good ventilation. It should not be stored with fruits or vegetables that produce ethylene gas as this can accelerate the ripening process and shorten the shelf life of the knol-khol.
5. Packaging: Knol-khol can be packaged in perforated plastic bags or in crates lined with damp cloths to maintain its moisture content. The packaging should be kept in a cool and well-ventilated area.
6. Transportation: During transportation, knol-khol should be handled carefully to avoid any damage or bruising. It should be transported in a cool and well-ventilated vehicle to maintain its quality.

By following these post-harvest management practices, the quality and shelf life of knol-khol can be maintained, which will increase its market value and profitability.

Yield

Early types' knobs may weigh 200-250g on average, whereas late kinds' knobs can weigh up to 1kg. In general, yields might range from 12 to 30 tons per acre.

Future Prospect

Knol-khol, also known as kohlrabi, is a vegetable crop that is widely grown and consumed in various parts of the world, particularly in Europe and Asia. The future prospect of production technology for knol-khol will depend on several factors, including market demand, technological advancements, and environmental considerations. One of the key factors that will shape the future of knol-khol production technology is market demand. As consumers become more health-conscious and interested in sustainable agriculture, there may be increased demand for knol-khol and other nutrient-rich vegetables. This could lead to greater investment in research and development aimed at improving the efficiency, yield, and quality of knol-khol production. Technological advancements, particularly in the areas of genetics, robotics, and precision agriculture, could also have a significant impact on the future of knol-khol production. For example, genetic engineering techniques could be used to develop knol-khol varieties with desirable traits such as disease resistance, drought tolerance, and improved nutritional content. Robotics and precision agriculture could also help farmers to optimize their production practices, reduce labor costs, and minimize environmental impact.

Environmental considerations will also be an important factor in the future of knol-khol production technology. As the effects of climate change become more pronounced, there may be greater emphasis on sustainable agricultural practices that reduce greenhouse gas emissions, conserve water resources, and minimize chemical inputs. This could lead to increased adoption of practices such as conservation tillage, crop rotation, and integrated pest management, which can help to improve soil health and reduce environmental impact.

References

Bose, T. K., Kabir, J., Maity, T. K., Parthasarathy, V. A. and Som, M. G. (2002). Vegetable Crops. Bhubani Mitra, Naya Prokash, 206, Bidhan Sarani, Calcutta. 1:345-392.

Mishra, P. P., Das, A. K. and Mishra, N. (2014). Effect of integrated nutrient management on yield, quality and economics of knolkhol (*Brassica oleracea* L. cv. *Gongylodes*). *Asian Journal of Horticulture*, *9*(2):382-385.

Kumar, M., Singh, V., Rana, D. K. and Shah, K. N. (2018). Effect of integrated nutrient management on various horticultural traits of knol-khol (*Brassica oleracea* var. *gongylodes*) cv. White Vienna under Garhwal Hills. *Journal of Pharmacognosy and Phytochemistry*, *7*(1):2285-2288.

Deelak, A., Sofi, M. A., Mir, G. M., Shahida, I. and Asmat, A. (2016). Effect of pollinator attractants on seed yield of knol khol. *Indian Journal of Ecology*, *43*(1):178-179.

Dhawan, A. K., Rajinder, P. and Chauhan, S. K. (2016). Natural Resource Management: Ecological Perspectives. *Indian Journal of Ecology*, *43*(2):603-610.

Silatar, P., Patel, G. S., Acharya, S. K. and Vadodaria, J. R. (2018). Performance of different varieties and plant spacing on growth and yield of knolkhol (*Brassica oleracea* var. *gongylodes*). *International Journal of Agricultural Sciences*, *8*(7), 1476-1479.

4

Advances in Production Technology of Brussels Sprouts

Suprava Biswal[1]* and Toko Yarin[2]

[1]Ph.D. Research Scholar, Department of Vegetable and Spice Crops, Uttar Banga Krishi Viswavidyalaya, Pundibari, Cooch Behar-736165, West Bengal

[2]Ph.D. Research Scholar, Department of Plantation Crops and Processing, Uttar Banga Krishi Viswavidyalaya, Pundibari, Cooch Behar-736165, West Bengal

**Corresponding author *Email: supravabiswalchc2@gmail.com*

Introduction

The 'Brussels sprouts' (*Brassica oleracea* L. var. *gemmifera*) coming under Brassicaceae family with chromosome number 2n=18 preferred by farmer for its edible green bud or sprouts (small heads that resemble miniature cabbages). It is a robust, slow-growing, long-season vegetable. The "sprouts" are formed in the leaf axils, beginning at the base of the stem and progressing upward.

Figure 4.1: Brussels sprout

Brussels sprouts are high in vitamin A and C, as well as folic acid and dietary fiber. It also has anticancer effects due to the presence of sulforaphane. Similarly, to other Brassicas, Brussels sprouts contain indole-3-carbinol, which inhibit cancer cell growth. The temperate regions in the hills have the greatest potential for commercial production of brussels sprouts, followed by coastal regions with prolonged cool falls. Brussels sprouts are mostly popularized in countries like UK, USA, Canada, Mexico, Germany, China and Netherland etc. Centre of origin of brussels sprouts is Belgium and wild ancestor i.e., broccoli, first grew in European subcontinent. Later on, it is introduced to the USA states and popularized by people.

Botany

Brussels sprouts is a cultivar group of the species *Brassica oleracea*, which also includes other vegetables like kale, cabbage, and broccoli. Brussels sprouts are biennial plants that grow best in cool temperatures. The plants can reach a height of 2 to 3 feet and have a spread of 1.5 to 2 feet. The leaves of the Brussels sprout plant are dark green and have a waxy texture. The plant produces small buds or sprouts that grow along the stem of the plant. These sprouts are typically 1 to 1.5 inches in diameter and are tightly packed together. The sprouts grow from the bottom of the stem to the top, and the lower sprouts mature first. Brussels sprouts require consistent moisture throughout their growing season, and they benefit from regular fertilization. The plants are also susceptible to a variety of pests and diseases, including aphids, cabbage loopers, and clubroot.

Taxonomy

Brussels sprouts are a member of the Brassicaceae family, also known as the cruciferous or mustard family. Within the Brassicaceae family, Brussels sprouts are classified as follows:

Rank	**Scientific Taxonomical Details**
Kingdom	Plantae (plants)
Division	Magnoliophyta (flowering plants)
Class	Magnoliopsida (dicotyledons)
Order	Brassicales
Family	Brassicaceae (mustard family)
Genus	*Brassica*
Species	*Oleracea*
Variety	Gemmifera

Varieties

There is not a single variety of brussels sprouts crop released in India.

1. **Hills Ideal-** This variety grown in north Indian plains and hilly zone with a medium-tall height (55-60 cm) having 45-50 sprouts (7-8 gm/sprout). Compact head with good flavor and usually takes 75-80days after transplanting. Average yield of this cultivar is 10-15 tons/ha.
2. **Rubine-** High yielding cultivar. Sprouts is also knowns as blowts show reddish color. Late maturing type so usually take 105 days for harvesting after transplanting.
3. **Prince Marvel-** Compact blowts with sweet in taste. Takes 90 days for harvesting after transplanting
4. **Jade Cross-** Japanese F1 hybrid, an early type, short stemmed and usually take 90 days for harvesting after transplanting.

5. **Oliver-** Widely popular hybrid cultivar of 1-1.25 ft. plant height. Green, coarse leaf type and smooth type. and
6. **Bubbles-** It is highly popular among people due to excellent flavor and having longer field staying capacity.
7. **Royal Marvel-** Extra early type cultivar and 2.5 cm size of each sprouts having round in shape and dark green in color. This variety has peculiar characteristic in nutritional point of view as it is rich source of vitamin and protein. Harvesting of small sprouts is very difficult due to evenly spaced in the stalk.

Climate requirement

Brussels sprouts are a cool-season crop that thrives in a cool, humid area. It has the highest production potential at temperatures ranging from 15 to 18°C. Sprouts grow best and enhance in quality when the weather is cool or even moderately frost.

Soil requirement

It can be grown effectively in a variety of soil types, but they thrive in well-drained loamy soil rich in organic matter. For early crops, sandy loams are ideal. It thrives in the pH range of 6.0 to 6.8.

Seed rate, land preparation and nursery management

A seed rate of 600-700 g is ideal for one ha. for crop production. Before sowing, seed should be treated with hot water (50°C) for 30 minutes against black rot and Apron 35 @ 2g/kg of seed against downy mildew. During land preparation disc plough is used to get a fine tilth, then harrow it one or two times. After that incorporation of well-decomposed FYM @ 25 t ha^{-1}. Organic manure or vermicompost or poultry manure application promotes plant growth, productivity, and field soil water retention capacity. The sterilizing of soil is achieved by spraying of nursery beds with formalin @ 1:49 approximately 15-20 days before seed sowing is effective for preventing fungal disease attack. Seed beds should be covered with polythene for a week after watering. The beds are then dug again and kept open for 5-6 days to prevent formalin from harming the seeds. The regular nursery-raising approach should be followed. Before watering, cover the beds with a suitable mulching material. During the early stage, 15-20 days after sowing, apply water using a water cane over the grass mulch, while later stage watering should be done through furrows. The mulch should be removed as soon as seed sprouts appear. The beds should have a roof to provide shade from the scorching sun and rain.

Transplanting of seedlings and spacing

Sowing of seed in the nursery for Brussels sprouts is early April, then transplant seedlings into the field in mid-May. After a month of seedling in the nursery, seedlings are ready to be transplanted in the field. A spacing of 45 X 70 cm^2 between plants as well as row to row should be followed. Turbin *et al.* (2014) achieved highest yield and uniform harvest of Brussel sprouts vegetable crop at plant spacing of 0.49 m^2/plant.

Nutrient requirement

The amount of manure and fertilizer needed for 'Brussels sprouts' is determined by the soil's fertility. During field preparation, application of 25 tons of well-rotten FYM helps in improvement of soil in terms of physical structure and texture. In addition, use of 100 kg N, 60 kg P_2O_5, and 40 kg K_2O for growth and development. Before transplantation, a half-dose of nitrogen and full doses of phosphorous and potassium should be applied. The remaining half dose of nitrogen should be top dressed twice, once after one month of transplanting and again when sprouts appear. Dolomite as lime should be applied to acid soils that are poor in Ca and Mg. Ali *et al.* (2022) applied potassium sulfate with organic manure i.e., poultry manure and observed better plant growth, small head growth and quality of brussels sprouts cole crops. Omar (2017) took two hybrid F_1 cultivar of brussels sprouts like Topline and Attwood and observed highest yield and growth factor at cow manure 20 ton/ha. among chicken and sheep manure.

Water management

The first irrigation occurs soon after transplanting, and further irrigation occurs as needed, based on soil and weather conditions. Irrigation can be repeated every 15 days interval. When sprouts grow, there should be enough moisture in the soil. Acharya *et al.* (2019) used low tunnel made up of spunbonded fabrics against cold and freezing conditions of brussel sprouts growth and development. They also concluded that less evapotranspiration occurred without affecting the brussels sprouts quality and achieved high water use efficiency due to low tunnel condition.

Weeding

Shallow frequent hoeing in the field with a 'khurpi or hoe' should be done to kill emerging weeds and give soil mulch. Plants should be slightly earthen up in the field four to five weeks after transplanting. Stomp 30 EC (Pendimethalin) 2.5lt/ha can be used one day before seedling transplanting under moist soil conditions for chemical weed control.

Topping

Brussels sprouts came in first as a major operation. The fully grown plants are cut off at the growth stage (topping). This is typically done to promote the growth of sprouts at the top of the plant and to ensure uniform development for once-over harvesting.

Harvesting and Yield

There are two methods are available for harvesting of brussels sprouts, firstly if all sprouts are mature at a time, then harvest of whole stalk/stem is necessary. Secondly, if all sprouts are not mature at a same period, then serial harvesting can be done by choosing at mature and marketable stage of individual sprouts. Harvesting of Brussels sprouts typically begins in late fall or early winter when the sprouts are mature and firm to the touch. The sprouts are typically harvested by cutting them off the stem with a sharp knife. Tender knobs should be harvested before they become fibrous. Harvesting occurs up to three times every week. When the leaves are tender, they are also used for green. The crop yields between 200 and 300 q ha^{-1}.

Post-harvest Management

Brussels sprouts are a nutritious and delicious vegetable crop that require careful post-harvest management to maintain their quality and freshness.

Harvest at the right time: Brussels sprouts should be harvested when they are firm and green. Wait until the sprouts are about 1-2 inches in diameter before harvesting. This ensures that they have reached their maximum size and flavor potential.

Remove any damaged or yellowing leaves: After harvesting, remove any leaves that are damaged, yellowing, or otherwise not healthy. This will help prevent the spread of disease and decay.

Sort and grade the sprouts: Sort the sprouts into different grades based on their size, color, and quality. This will help ensure that you are selling or storing only the best sprouts.

Chill the sprouts: Brussels sprouts should be cooled as quickly as possible after harvesting. The ideal storage temperature for Brussels sprouts is between 32-36°F (0-2°C). This will help to maintain their freshness and quality.

Packaging: Pack the sprouts in perforated plastic bags or in ventilated containers to allow air circulation.

Transport and storage: During transport and storage, keep the sprouts away from ethylene-producing fruits and vegetables, as they can cause the sprouts to ripen and deteriorate more quickly.

Keep humidity levels in check: Maintaining humidity levels around 95% will help prevent the sprouts from drying out.

By following these steps, you can ensure that your Brussels sprouts remain fresh and nutritious for a longer period, making them more attractive to buyers and consumers.

Future Prospect

Brussels sprouts are a popular vegetable that have been cultivated for centuries, and their production technology is likely to continue to evolve in the future. Here are some potential future prospects for the production technology of Brussels sprouts:

Sustainable production: With increasing awareness about the impact of agriculture on the environment, there is likely to be a greater focus on sustainable production methods for Brussels sprouts. This could include the use of organic fertilizers, precision farming techniques to reduce waste and optimize resource use, and the use of integrated pest management strategies to reduce reliance on pesticides.

Biotechnology: Advances in biotechnology could lead to the development of new Brussels sprout varieties that are more resistant to pests and diseases, have higher yields, and are more nutritious. Genetic modification could also be used to produce Brussels sprouts that are more resilient to climate change.

Automation: Automation is already being used in many aspects of agriculture, and this trend is likely to continue in the future. This could include the use of robotics and artificial intelligence to optimize planting, harvesting, and processing of Brussels sprouts.

Vertical farming: As urbanization continues to increase, there may be a greater demand for locally-grown produce, which could lead to increased adoption of vertical farming techniques for Brussels sprouts. Vertical farming allows for high-density production in a controlled environment, with minimal water use and no soil required.

Overall, the future of Brussels sprouts production technology is likely to be shaped by a range of factors, including sustainability concerns, technological advancements, and changing consumer preferences.

References

Acharya T. P., Welbaum G. E. and Arancibia, R. A. (2019). Low tunnels reduce irrigation water needs and increase growth, yield, and water-use efficiency in Brussels sprouts production. *Hort Science*, 54(3):470-475.

Kumar A, Choudhary AK and Rahi S. (2014). Scientific cultivation of brussels sprouts (*Brassica oleracea* L. *var. gemmifera*). Advances in Vegetable Agronomy. Indian Agricultural Research Institute, New Delhi. (1):92-95.

Samal Jalal Omar. (2017). Effect of Different Manures on the Growth, Yield and Quality of Two Brussels Sprouts (*Brassica oleracea* var. *gemmifera* L.) Hybrids. *Euphrates Journal of Agriculture Science*. 9(4):26- 41.

Sarmad Salim Ali, Ali Ibadi Manea and Bassim Haleem Kshash. (2022). Effect of Foliar Application with Potassium Sulphate and Organic Fertilizer on Growth Parameters and Yield of Brussels Sprouts (*Brassica oleracea* L. *var. gemmifera* Zenk). IOP *Conference Series: Earth and Environmental Science*.1-5.

Turbin, V. A., Sokolov, A. S., Kosterna-Kelle, E. A. and Rosa, R. S. (2014). Effect of plant density on the growth, development and yield of brussels sprouts (*Brassica oleracea* L. *var. gemmifera* L.). *Acta agrobotanica*, 67(4):51-58.

5

Advances in Production Technology of Broccoli

Toko Yarin[1] and Suprava Biswal[2*]

[1]Ph.D. Research Scholar, Department of Plantation Crops and Processing, Uttar Banga Krishi Viswavidyalaya, Pundibari, Cooch Behar-736165, West Bengal

[2]Ph.D. Research Scholar, Department of Vegetable and Spice Crops, Uttar Banga Krishi Viswavidyalaya, Pundibari, Cooch Behar-736165, West Bengal*

**Email: supravabiswalchc2@gmail.com*

Introduction

Figure 5.1: Broccoli

Broccoli (*Brassica oleracea* L. var. *italica*) having chromosome number 2n=2x=18 is a native of Mediterranean region. The broccoli was developed in eastern Mediterranean region and Italy is the main diversification which was later introduced into England about 1720. It is a cool season crop which produces fleshy green flower heads arranged on branches sprouting from thick edible stalk in such way that it looks like a miniature bonsai tree. It has the same family as cabbage and cauliflower. Even though it resembles cauliflower by forming head like structure but it has longer flower stalks with loose large green terminal head (10 to 20 cm). In Italy broccoli is called as broccoli which means "the flowering crest of a cabbage." Due to the presence of bountiful nutritious properties, it is also known as "Crown Jewel of Nutrition". It is a great source of vitamins A, B and C and minerals like iron, fibre, calcium, zinc, magnesium etc. Apart from the minerals and vitamin it also a rich source of riboflavin, thiamine and niacin. Due to its high content of bioactive compounds and health benefits properties, it is considering as important food for prevention of chronic diseases like cancer (breast and prostate cancer) and cardiovascular and carcinogenic pathologies. Broccoli is mostly consumed as a cooked vegetable and in salad form or just by slightly steam. It is mostly marketed as fresh and frozen vegetables especially in USA. During early 90 centuries, knowing its high nutritional value the crops have been introduced to many other countries particularly among the city dwellers. China is the largest producer country which followed by India with annual production of 6.7 million tons while the US ranks third position in the world (Kumar *et al.,* 2014 and Tejaswini *et al.,* 2018).

Botany and Taxonomy

Broccoli is the young shoots that develop during spring on some species of the genus brassica. Broccoli is originally, originated from natural crossing between ancestral broccolis (*Brassica cretica* ssp.nive) of the wild annual with white cymose flowers and the wild biennial types after that by human selection thousands year ago. So, due to its open flower bud as compared to cauliflowers prefloral fleshy apical meristem, it is believed that broccoli must have preceded cauliflower in the evolution of species. The plant consists of a short stem and rosette of leaves in early developmental stage with internodes elongation and at certain point the terminal axis proliferates into numerous branches and form small head on the axil of leaves (Bose *et al.*, 2002). The terminal buds are loose, purple and green colours depending upon the varieties. The sprouts in the axils of leaves develop profusely, especially after the removal of the terminal head. The plant is generally regard as an annual and a period of cold temperature, is not required to induce flower bud initiation. Although flower is hastened by exposure to low temperatures (Sing and Nath, 2012).

Improved Cultivars

Broccoli cultivars are mainly categorized according to their shape and head size, maturation time such as early, medium and late cultivars, secondary shoots production and disease resistance. There are mainly three types of broccolis i.e., based on their head colours as well as days of crop maturity such as Calabrese, sprouting and purple cauliflower broccoli.

Calabrese Broccoli

It is the most common type of broccoli that is commonly cultivated by the farmers. It is a cool season crop with a thick stalks and large green head. It is simply called as 'broccoli."

Sprouting Broccoli

This is a white and purple colour broccoli that produce from accumulation of many long thin stalks with numerous numbers of heads.

Purple Cauliflower or Violet Cauliflower

These colorful broccolis are mostly grown by the Europe and North America. The head shape of this broccoli resembles to cauliflower but has many tiny flower buds as compared to cauliflower. Following are the characteristic of some improved varieties accordance of their maturation period.

Early season type

1. **Blue wind**: It is an extra early F1 hybrid variety which matured in 49-55 days. It produces a big and compact head in 60 days after transplanting and produces good side shoot for an extensive harvesting. The top of plant is surrounded with tender and bluish-green leaves which is use for steam or braise like kale.
2. **Belstar**: It is a hybrid between Green Comet and Blue Lake 274 and is mostly grown in California, Arizona and Mexico. The plant grows to a height of 2 to 3 feet tall and produces yellow flowers. It takes 65 to 75 days to mature after transplanting.
3. **De Cicco:** It is an open pollinated variety which is known for its large and green florets that arranged in a tight cluster. It is also known as Italian Sprouting or Tuscany Kale. It can be harvested in 60 to 75 days after transplanting.
4. **Arcadia:** It is a hybrid variety which produces uniform and purplish-green heads. It is tolerant to sever cold, resistant to head rot, brown head and downy mildew. It is ready to harvested in 63 to 68 days after transplanting.
5. **Gypsy:** This variety is also known as wild broccoli or Calabrian broccoli. It is a very early season variety which thrives well also in poor soil and gives good productivity. It grows nicely round domes green heads with medium to tiny bead sizes that are heat tolerant.
6. **Atlantic:** It is also a hybrid type that yields a lot of flavorful, highly productive florets and stems with deep green hues. Regular harvesting promotes more floret development.
7. **Amadeu:** This variety produces large, tender heads with an excellent flavor. It is a season crop that grows best in temperate climates but prefers full sun and can tolerate some shade. The plant grows to a height of 2 to 3 feet and produces large, green heads that weigh 1 to 2 pounds each. The heads are ready to harvest in 60-80 days after planting.

Mid-season type

1. **Punjab broccoli-1:** This variety was developed by PAU Ludhiana in 1996(Anonymous, 2012). The plant is attractive and has bluish tinges leaves with compact and succulent sprouts. It is ready to harvest in 65 days after transplanting and yield 175 q/ ha.

2. **Palam Samridhi:** It is a HYV and the weight of each terminal head is about 300-400 gm. This green sprouting broccoli is tender, fresh with good flavour, large loose heads and is suitable for salad, cooked vegetable and pickled. It gives 150-200 q/ha of tender heads (Kalia, 2002).
3. **Pusa KTS 1:** The plant height is 65 cm to 70 cm and produces compact heads with dark green waxy foliage. The head is about 6.0 cm to 15.4 cm in size and 350 to 450 gm weight. It is ready for harvesting in 90-105 days after transplanting.
4. **Green Sprouting Medium:** It takes about 100 days to maturity after transplanting.
5. **Pusa Broccoli Kt. Sel. 1:** This variety is released from IARI, Regional Station Katrain (Kullu Valley). Harvesting is done 85 to 95 days after transplanting when the plant attains 40 to 50 cm height. The solid light green head weigh approximately 250-400g.

Late season type

1. **Waltham 29:** It is an open pollinated variety with high productivity and produces more side shoots after the harvesting of main head. The variety is tolerant to very cold season and can be grown for early spring and late fall production.
2. **Palam Vichitra:** It is a medium-sized heading type broccoli with a stalk that has dark green leaves and a stem with a purple tint. The dense, purplish heads are loaded with vitamins and minerals. Average yield is 225 q/ha.
3. **Palam Haritika:** It is a sprouting broccoli and the main head developed deep inside the dark green straight leaves. It takes 150 days to produce and is free of yellow eyes despite its late maturity (anonymous, 2012). The yield produced on average is 200–250 q/ha.
4. **Palam Kanchan:** It is a yellow-coloured heading type broccoli that has large green leaves and white midrib. Harvesting is done in 140 days after transplanting and obtain 250-300q/ha yield.
5. **Coastal Atlantic:** It is cultivated as early and late season crop. Plant is sensitive to buttoning.

Soil and climate

Broccoli can be cultivated in a wide range of soils but thrives best in well drained, deep loamy soil with sufficient organic matter. For a vigorous growth it requires moist soil as dry soil produces more fibrous shoots. The optimum pH of the soil should be 6.0 to 6.5 for better cultivation.

It is a cool season crops like the any other Cole crops and thrives best at average temperature ranges between 20°C to 25°C while 15°C to 20°C is the best for heading stage. If the temperature is below the optimum range, then there is a delay in maturity which ultimately led to the small sprouts (Kumar *et al.*, 2014). High temperature gives rise bracts during developmental stage of head and head becomes loose.

Cultural practices

Nursery management

For broccoli cultivation 400–500 grams of seed are sufficient for one hectare. Raised the nursery bed at 15 cm from the ground and prepared the land with well decompose cow dung manure along with SSP @ 50 to 60 grams per sqm per bed. For the sterilization of soil, drench the bed with 4.0% formaldehyde and then cover the bed with polythene sheet for 5-7 days.

After this, treat the seed with 1gm carbendazim or 2.5 grams thiram per kilogram of seeds. Sow the treated seeds at shallow depth at 5.0 cm apart, 1-1.5 cm depth and covered finely sieved vermicompost. The beds should be covered with straw and give light irrigation with fountains till the seeds germinate. It is necessary to cover the nursery beds with weed or polythene sheets to protect from rain water.

Planting time

Sowing time of broccoli is depending on the variety, season, and location. Generally sowing is done from mid-July to first week of August for early varieties in mid hilly region and during March- April whereas, in plains sowing is done in October (Hajra, 2019).

Field Preparation

The land should be prepared well in advance with repeated ploughing (4-5 ploughing) to fine tilth and remove all the stubbles and weed to make the field leveled. Then add well rotten manure 10 to 15 tons per hectare and mixed well in the soil.

Transplanting

Before transplanting hardening of the seedlings should be done by withholding water prior 4 to 6 days. Plant having 4-6 leaves at a height of 6-8 cm is ideal for transplanting. Planting should be done on ridges to avoid water logging problem. Planting is varying according the varieties as well as climate and soil condition. So, the planting should be done at a spacing of 45 cm × 45 cm at 3 to 4 cm deep.

Nutrient Management

Depending upon the soil fertility status the recommended doses of manure and fertilizer is apply in broccoli field. The recommended dose of NPK is 100-125 kg N, 75 kg P_2O_5 and 50 kg K_2O per hectare. Apply 20 tonnes FYM at the time of field preparation with full doses of P and K and half dose of N whereas, the rest dose of N should be applied as a top dressing in two equal splits viz. one month of after transplanting and last at the time head formation.

Irrigation

Broccoli requires enough moisture for uniform and continuous growth. So, irrigate immediately after transplanting. Generally, irrigation is given at an interval of 7 to 8 days in summer season and 10 to 15 days during winter season on the basis of the soil and climate condition.

Interculture operation

During the early stage of crop, field should be kept weed free and maintain proper moisture condition of the soil to promote constant growth. The weeding should be done with kurpi or hoe at a depth not more than 5 to 6 cm and if necessary, earthing should also be done.

Plant Protection

There are attack of insect, pest and disease infestation is quite similar to other cole crops like cabbage, cauliflower etc.

Harvesting and Yield

Harvesting should be done immediately after the bud cluster are green and compact as well as attained marketable size i.e., 10 to 15 cm by cutting the stems with a sharp knife. Delayed of harvesting may cause yellowing and loose bud cluster. After harvesting of main tip, other sub sprouts are again ready for harvesting after 10-12 days and it should be harvested before the bud's bloom at 8 to 10 cm long stalks. The yield of the broccoli depends upon the varieties and maturity period of the plant. Generally open pollinated varieties produced 10-15 t/ha while hybrid varieties yield 18-20 t/ha.

Postharvest handling and storage

After harvesting sprouts should be marketed as soon as possible as they cannot be stored for a long time. Before sending to the market, the heads should be immediately sorted, graded then packed in basket. They should be stored in cool temperature at 4°C and then packed with ice in crates and stored in refrigerator as high-rate respiration deteriorate its quality. At 4°C temperature it can be store for 7 to 10 days. Storing of broccoli florets in ethanol vapour enriched atmosphere at moderate temperature (Corcuff *et al.,* 1996). Fan and Mattheis, 2002 suggested that 1-methylcyclopropene treatment inhibits yellowing of broccoli and reduces its respiration rate even when the broccoli is exposed to ethylene as ethylene is the main factor of yellowing in broccoli.

Future prospect

Due to its reputation as a nutrient-rich food, broccoli is in high demand, and its market share is growing daily. However, production is not keeping up with customer demand. People are more attracted to broccoli and want to include it in their diet plans because it has some anti-cancer and nutritional properties. This element supports small marginal farmers and kitchen gardeners to adopt its cultivation and market production. There is still a demand for broccoli seed in the Indian market even though India is quickly overtaking China as the world's second-largest supplier of the vegetable. Fresh broccoli is still in limited supply in the markets as a result of poor storage life caused by a high respiration rate. So, there is a still need of proper manual for broccoli cultivation as well as good varieties and storage technique amongst the small margin famer. Therefore, in the upcoming years, these variables may accelerate the growth of broccoli in all markets throughout India.

References

Anonymous (2012) https://www.apnikheti.com/en/pn/agriculture/horticulture/vegetable-crops/broccoli

Bose, T. K., Kabir, J., Maity, T. K., Parthasarathy, V. A. and Som, M. G. (2002). Vegetable Crops. Bhubani Mitra, Naya Prokash, 206, Bidhan Sarani, Calcutta. 1:345-392.

Corcuff, R., Arul, J., Hamza, F., Castaigne, F. and Makhlouf, J (1996). *Post harvest Bio. Tech.*, 7:219-229.

Fan, XT and Mattheis, JP (2002). *Hort. Science*, **35**:885-887.

Hazra, P. (2019). Vegetable Science & Technology. New India publishing agency, New Delhi, 1:272-274.

Kalia, P. (2002) (Indian Agricultural Research Institute, New Delhi (India). Division of Vegetable Crops). Palam Samridhi: a new variety of sprouting broccoli. *Indian Horticulture (India)*, **47**(1): 24-25.

Kumar, A., Choudhary, A. K., and Rahi, S. (2014). Scientific Cultivation of Broccoli (*Brassica oleracea* L. var. *italica)*. Advances in Vegetable Agronomy. Indian Agricultural Research Institute,1: 87-91.

Sing, D. N. and Nath, V. (2012). Winter vegetables- Advances and developments. Satish serial publishing house, 1: 415-436.

Tejaswini T, Varma LR, Verma P, Prajapathi RI and Vani FB (2018). Performance of Different Varieties with Respect to Growth, Yield and Quality of Broccoli (*Brassica oleracea var. italica* L.) under North Gujarat Conditions. *Int. J. Curr. Microbiol. App. Sci.* 7(06): 690-698.

6

Advances in Production Technology of Kale

Suprava Biswal[1]* and Toko Yarin[2]

[1] Ph.D. Research Scholar, Department of Vegetable and Spice Crops, Uttar Banga Krishi Viswavidyalaya, Pundibari, Cooch Behar-736165, West Bengal

[2] Ph.D. Research Scholar, Department of Plantation Crops and Processing, Uttar Banga Krishi Viswavidyalaya, Pundibari, Cooch Behar-736165, West Bengal

**Corresponding author Email: supravabiswalchc2@gmail.com*

Introduction

Kale (*Brassica oleracea* var. *acephala*) belongs to family Cruciferae or Brassicaceae is a under exploited vegetable crop. It is introduced in 19th century.

Figure 1.1: Kale

Economic Importance and Uses

Its edible part is leaves and contains amazing source of amino acids, vitamins and minerals which has many medical advantages. It helps in lowering of blood cholesterol level which helps in lessening the heart illness. Kale leaves is also rich source of beta carotene, ascorbic acid and phylloquinone. Kale is a nutritious vegetable that is rich in vitamins A, C, and K, as well as calcium and iron. It is also a good source of antioxidants and fiber, making it a popular choice for health-conscious consumers. Kale is not so popular crop in India therefore it confined to very few states of Northern India like HP, UP, J & K, Pb and Haryana.

Botany

Kale is a leafy vegetable that belongs to the Brassicaceae family, which also includes broccoli, cauliflower, and cabbage. It is a biennial plant that is grown as an annual for its edible leaves. Kale has a compact rosette of leaves that are tightly curled or fringed at the edges. The leaves are usually green, but they can also be red, purple, or blue. The plant can grow up to 1 meter tall and has a deep taproot system that can reach up to 1.5 meters in depth. Kale is propagated through seeds, which are sown directly into the soil or transplanted from seedlings.

Taxonomy

Kale is a leafy green vegetable that belongs to the plant family Brassicaceae (also known as cruciferae). Its scientific name is *Brassica oleracea* var. *acephala*. Kale is a member of the same species as other vegetables such as broccoli, cauliflower, and Brussels sprouts. Within the species *Brassica oleracea*, there are several cultivar groups, each with their own distinct characteristics.

1. Kale is a part of the acephala group, which includes other leafy vegetables like collard greens and spring greens. The acephala group is characterized by their loose, non-heading leaves that grow from a central stem.
2. There are several varieties of kale within the acephala group, including curly kale, Tuscan kale (also known as Lacinato or dinosaur kale), and Russian kale. Each variety has its own unique appearance and flavor profile. Overall, the taxonomy of kale is as follows:

Rank	Scientific Taxonomical Details
Kingdom	Plantae
Clade	Tracheophytes
Division	Magnoliophyta
Class	Magnoliopsida
Order	Brassicales
Family	Brassicaceae
Genus	*Brassica*
Species	*Oleracea*
Variety	Acephala

Climatic requirement

The vegetable is the hardiest vegetable among all the cole crops to withstand in a heavy frost condition. Fundamentally, for better growth and development it requires ambient temperature of 10-15°C with a cold environment.

Soil requirement

The yield prefers a well-drained sandy loam soil rich in organic matter. It can grow well in saline soil condition without compromising on yield. A soil pH of 5.5-6.5 (slightly acidic) will result in a high yield.

Cultivars of Kale

Generally, three important cultivars are mostly adopted by farmer namely dwarf type with a height of below 40cm, medium type with a height of 40-80cm and tall type with a height of more than 80cm. 'Karam Saag' is a cultivar of medium type kale are mostly grown in J & K region. Tall type varieties are mostly shed their leaves at early stages of growth example cheeseburger market. Other types of cultivars namely Siberian and Scottish are mostly cultivated in developing countries.

Order Type of Kale Leaf

1. Curly leaf (Scots kale, blue twisted kale)
2. Plain leaf (level leaf types like Red Russian and White Russian Kale)
3. Leaf and skewer, or padded type leaf (a combination of a wavy and plain leaf)
4. Ornamental (less acceptable and harder leaves)

Sowing of Seed

The most well-known way of dissemination is by seeds. Seeds are typically placed in nursery beds to raise seedlings before being transported to the main continent.

Preparation of Land and Transplanting

The field should be prepared until it reaches a fine tilth level by allowing a few deep ploughings, which also aid in the removal of any weeds from the mainland. Enhancing poor topsoil with totally degraded FYM will extend the dirt ripeness and therefore the wonderful crop. For a one-hectare plot, a standard seed rate of 400-500gms is considered appropriate. In general, the months of Aug through Oct are considered to be the best sowing months for North Indian conditions. A month and a half old nursery elevated saplings should be transferred and transplanted at a 45 cm (R-R) X 30 cm (P-P) spacing. Dividing can be reduced for late planting.

Weed management and Irrigation

Weed control is a vital and critical stage in growing this vegetable. To keep the field weed-free, gentle hoeing and weeding should be done on a regular basis. In general, 3-4 hoeing and wedding are sufficient, and after the leaves cover the topsoil, there is no need for hoeing. Following field relocation, the starting water system should be provided, and subsequent irrigation should be performed every three weeks. Kale requires regular watering, especially during dry spells, but it should not be overwatered as this can lead to fungal diseases. It is also susceptible to a variety of pests and diseases, including aphids, cabbage worms, and powdery mildew.

Nutrient Management

For good and quality output, adding 25 tons FYM/hectare to the field at the time of preparation of land is often advised. Basic fertilizer of phosphorous and potassium @ 75kg/ha. full doses applied before the transplanting, while nitrogen-based fertilizer @ 150kg/ha. applied at three split doses 1/3rd part should be given at transplanting time, then again one part after several recommended months after transplanting and last part should be given at 3 weeks before first harvesting. Haile and Ayalew (2018) observed that application of organic fertilizer like bio slurry with chemical fertilizers at different concentrations gave highest leaf fresh weight and quality of leafy vegetables i.e., kale. Ngugi *et al* (2021) concluded that application of silicon gave better result in terms of leaf growth in leafy type vegetable like kale, Amaranth and spinach in cadmium or hard metal retains soil. Bernardi and Sala (2021) also concluded that kale is a highly source of vitamins, minerals and nutrients so use of organic fertilizer like organo mineral fertilizer complex helped in improvement of productivity of hybrid variety Darkibor of curly kale. Helaly *et al* (2022) treated the seeds of kale for 3hrs with 3 bacterial strains like AP-28 of *Pseudomonas koreensis*, AP-51 of *Bacillus cereus* and AP-29 of *Ralstoniapickettii*. They observed highest vegetative growth and quality in kale.

Harvesting

The leaves begin to grow, and harvesting begins in Nov and continues until Jan. To achieve a high quality, it should be harvested within the optimum vegetative season. Kale is typically harvested when the leaves are young and tender, usually within 50 to 70 days after sowing. The leaves can be harvested individually, or the entire plant can be cut at the base.

Yield

Kale leaves have good taste grown in chilled temperature. A standard yield of 100 to 200 quintals per hectare may be obtained by using good agricultural practices and suitable variety according to climate.

Post-harvest Management

Post-harvest management of kale, a leafy green vegetable, is an important step in ensuring the quality and freshness of the vegetable from the time of harvest to the point of sale or consumption. The main steps involved in the post-harvest management of kale include:

Cooling: As soon as the kale has been harvested, it should be cooled to prevent spoilage and maintain its freshness. This can be done by storing it in a cool place or by using refrigeration.

Grading: After cooling, the kale should be graded based on its size, shape, and appearance. Leaves that are uniform in size and shape will have a longer shelf life and be more marketable.

Packaging: The graded kale should be packaged in moisture-proof containers, such as plastic bags or containers, to prevent moisture loss and spoilage. The containers should be properly labeled with the type of kale and the date of harvest.

Transport: The packaged kale should be transported to the point of sale or storage in a manner that protects it from damage and maintains its freshness. This may involve using refrigerated transport or insulated containers.

Storage: If the kale is not going to be sold immediately, it should be stored in a cool, dry place that is free from temperature fluctuations. The kale should also be protected from moisture and light to maintain its freshness and quality. By following these post-harvest management steps, you can ensure the quality and freshness of your kale and maximize its shelf life.

Future Prospect

The future prospects of kale, a leafy green vegetable, are promising. Kale is widely recognized as a superfood due to its high nutrient content, including vitamins, minerals, and antioxidants. This has led to increased demand for kale in both traditional and emerging markets. In addition, the popularity of plant-based diets and the growing trend toward healthier eating habits have also contributed to the rising demand for kale. The increasing demand for kale is also driving the development of new and innovative products, such as kale chips, kale powder, and kale juice, which are being introduced to the market to meet consumer demand. This presents opportunities for growers, processors, and marketers to expand their businesses and tap into the growing market for kale. In addition, advances in agricultural technologies, such as precision agriculture and controlled environment agriculture, are making it possible to grow kale in a more efficient and sustainable manner. This, combined with the growing demand for locally grown produce, presents opportunities for growers to expand their production of kale and meet the needs of local markets. To meet the growing demand for kale, crop improvement is an important aspect. This includes breeding for disease and pest resistance, improving irrigation and fertilization, implementing effective pest management strategies, and proper post-harvest handling. With continued research and improvement, kale has the potential to become an even more valuable and widely grown crop in the future, providing consumers with a healthy and sustainable food option.

References

Bernardi,V.,M. R., Pimenta, D. M., Levrero, G. R., Forti, V. A., de Medeiros, S. D., Ceccato-Antonini, S. R. and Sala, F. C. (2021). Yield and quality of curly kale grown using organic fertilizers. *Horticultura Brasileira, 39*:112-121.

Bose, T. K., Kabir, J., Maity, T. K., Parthasarathy, V. A. and Som, M. G. (2002). Vegetable Crops. Bhubani Mitra, Naya Prokash, 206, Bidhan Sarani, Calcutta. 1:345-392.

Haile, A.and Ayalew, T. (2018). Comparative study on the effect of bio-slurry and inorganic N-fertilizer on growth and yield of kale (*Brassica oleracea* L.). *African Journal of Plant Science, 12*(4):81-87.

Helaly, A. A., Mady, E., Salem, E. A. and Randhir, T. O. (2022). Stimulatory effects of growth-promoting bacteria on growth, nutritional composition, and yield of kale plants. *Journal of Plant Nutrition, 45*(16):2465-2477.

Ngugi, M. M., Gitari, H. I., Muii, C. and Gweyi-Onyango, J. P. (2022). Cadmium mobility, uptake, and accumulation in spinach, kale, and amaranths vegetables as influenced by silicon fertilization. *Bioremediation Journal, 26*(2):113-127.

7

Advances in Production Technology of Chinese Cabbage

Toko Yarin[1] and Suprava Biswal[2*]

[1]Ph.D. Research Scholar, Department of Plantation Crops and Processing, Uttar Banga Krishi Viswavidyalaya, Pundibari, Cooch Behar-736165, West Bengal

[2]Ph.D. Research Scholar, Department of Vegetable and Spice Crops, Uttar Banga Krishi Viswavidyalaya, Pundibari, Cooch Behar-736165, West Bengal*

[*]Email: supravabiswalchc2@gmail.com

Introduction

Chinese cabbage (*Brassica compestris var. pekinensis* i.e., heading type and *Brassica compestris var. chinensis* i.e., non-heading type) with chromosome number 2n=20 is a green leafy vegetable that has come under the same family Cruciferae as cabbage. It is also called as celery cabbage, Bok choy, Peking cabbage, snow cabbage, napa cabbage etc. according to the places. It is originated from China and later adopted to other countries, mostly in Asian countries. Even though not being a member of cole group, it is regarded as one because of the similarity in cultural practices, breeding systems, same growing season and many other features that resembles apart from being a *brassica* vegetable. The best planting season for Chinese cabbages is in early spring to harvest as a late spring and after mid-summer for a fall crop. The inner leaves of the plant are light yellow colour with long (oblong) shaped head which is covered by compact wrinkled, profuse and light green leaves. Generally, the crop has high nutritional value due to which its demand is increasing tremendously. The potassium content of the crop is high with low fat content. It is also rich in vitamin C, fiber, folate as well as beta carotene and calcium are also more as compared to the cabbage due its leaves colour.

Botany and Taxonomy

Chinese cabbage is technically a biennial plant but is grown as annual crop for the seed production. It has two species of cabbage that belong to the family Brassicaceae i.e., *Brassica rapa pekinensis* (napa cabbage) and *Brassica rapa chinensis* (bok- choy). It is a fast-growing vegetable which can be harvested within three to four weeks after sowing. The plant has a light green leaf with white petiole and is more delicate than the cabbage. The shape and size of the head is differed according to their varieties and weighed between 1.4 and 4.5 kg. The inside of leaves is usually creamy white colour whereas the outer side varies from dark to light green. The bulbous base form a solid white 'butt' to the harvested head's base

According to the compact arrangement of internal leaves of head, the Chinese cabbage is divided into the two types :

1. **Heading types**- It is of two types i.e., Michihili and Napa; While the Napa has a small barrel-shaped head whereas Michihili is a tall cylindrical or tapered head type. It can grow upto a height of 25 cm to 51 cm with a thickness of 15-20 cm. These can be cultivated similarly to cabbage, but they have a mild flavour as compared to the cabbage. These heads are compact when they are clipped. Harvesting is done by cutting off the entire head.

Figure 1.1: Napa cabbage

2. **Loose-heading types (pac choi, bok choy and pei tsai)-** Both pac choi and pei tsai have loose, open heads with dark green leaves and white stems that bear a resemblance to celery. The plant grows upto a height of 15 cm to 60 cm tall and spread outwards by 15 – 46 cm. These can tolerate heat, but if the temperature rises sharply from cold to very warm, they will flee. These types are planted at two to three weeks intervals for a continual harvest.

Figure 1.2: Bok choy

Important Cultivars

Li (1981) has classified Chinese cabbage into four varieties:

1. **Loose leaved variety (var. *dissoluta*):** It is a under developed terminal buds with no leaf bud, rosette leaves are oblanceolate with spreading or erect. It is mostly grown during spring and summer in China.
2. **Semi heading variety (var. *imfareta*):** This type of variety is generally grown during the autumn season and plants are tall and large with terminal bud covering with well-developed outer layer of leaves as well as hollow head formation.
3. **Fluffy topped heading variety (var. *laxa*):** Plants are modest, with rosette leaves that are obovate and spread outward and a terminal bud that is fully formed and resembles a solid head. often grows in the late summer and fall.
4. **Heading variety (var. *cephalata*):** Well-developed terminal bud forming a compact leaf head tips that nearly overlap on top. For autumn season harvesting, it is grown as early crop and for winter as late crop. Based on its morphological and ecological characteristic, the variety has been categorized into three types viz. cylindrical (f. cylindrical Li), flat-topped (f. depressa Li) and ovate (f. ovata Li,)

Improved varieties

Palampur Green

- This a late flowering type plant having green leave with tender and creamy stem.
- It gives 5 -6 cuttings and first harvesting is done 23-30 days after planting.
- Harvesting is done in 15 days interval after the first harvesting.

Pusa sag

- It is a hybrid of Wongbok (Suttons) x Turnip.
- It tastes similar as the other sag.

Solan Band Sarson

- It is a heading type variety with elongated and compact heads weighing 700-1100g with 6-9 outer leaves.
- Harvested in 120 days after planting when the colour of the leaves turns light and golden.

- It is mostly used for salad and cooking purpose and can be transported to long distance for a marketing purpose.
- It yields an average of 400q/ha.

Solan Selection

- The leaves are tender and petioles are light green colour.
- The yield obtained from this variety is 150-190q/ha.

Chinese Sarson No. 1

- The leaves are light greens, broad leaves and semi-erect with white succulent tender mid rib.
- First cutting is done 30 days after transplanting.
- Rejuvenating in nature and can be obtained 6 to 8 cuttings.
- Yield is 40q/ha.
- Resistance to disease *Alternaria* leaf spot.

Area and production

Approximately 260 000 acres, or 27% of the entire vegetable producing area, was under cultivation in China in 1982. In 1988, over 32 000 acres, or 5% of the country's total vegetable producing area, were planted with this crop in Japan. During the year 1986, Korea's 13% of the nation production was vegetable production with 48 000 acres. White cabbage had the second-largest production area in Taiwan in 1988 with 8000 hectares and the total 4% was vegetable production. In 1986, this crop and Pak choi were planted on about 8000 acres in Thailand. Due to the availability of cultivars suited to tropical climates, output in South-East Asia has significantly increased in recent years (Kuo & Toxopeus, 2016).

Climate and Soil

It is also a cool season crop like any other cole crops but in temperate region it is grown during the summer, autumn and spring season under protected condition. Crop yield decrease when there is fluctuation in air temperature i.e., either too high or too low. The highest yield is obtained at optimum temperature ranges between 12-22°C but temperature above 25°C prevent the formation of marketable heads of most of the temperate cultivars. Chinese cabbage required enough organic manure and moisture for its growth. It also grows well on deep fertile clayey loamy soil with pH of 6.0-7.8.

Cultivation Practices

Chinese cabbage is propagated same as the other cole crops through seed. Seeds are raised in nursery bed for seedling purpose or can be directly sown at main field. But sometimes the problem that faced in direct sowing seed in field is that it may not give enough time for seedlings to grow before the summer days to develop seed stalk and bolting (Anonymous). In plains the best time of planting Chinese cabbage is mid September to end of November and in mid hills April to July. Depending upon the season and the varieties it may take 50 to 100 days for the crop to mature which is less as compared to the cabbage and cauliflower (Anonymous). Apart from the growing period of chinses cabbage most of the cultural practices are same as the cabbage and cauliflower.

The seed rate required for direct sowing is 2-2.5kg/ha and for transplanting @500-750g/ha. The planting should be done at a spacing of 30-40 x 30-40 cm for early variety whereas for late maturing variety at a spacing 70 cm x 55 cm (Anonymous).

Nutrient management

Chinese cabbage thrives best with the application of recommended doses of manures and fertilizer. During the field preparation time apply 15-20 t/ha well decomposed FYM along with the recommended doses of NPK i.e., 200-220: 100-150:200-250 Kg/ha (Letard,1984). The application of nitrogen, phosphorus and potash varies according to the region, soil types and different varieties. During the time of transplanting half dose nitrogen and full dose of both potash and phosphorus should apply whereas remaining dose of nitrogen should apply 30 days after transplanting.

Irrigation

Furrow method of irrigation is followed for the Chinese cabbage and the plant should be irrigated immediately after transplanting whereas, next irrigation should be given 0-15 days interval depending on the soil condition. At the time head maturity stage heavy irrigation should be avoid.

Intercultural operation

Weeding is done by using hand hoeing, machine and applying herbicide/ weedicides. Mulching is done to retain the moisture content as well as to control the growth of weeds. Earthing up is also done 2 months after planting.

Physiological disorder

- **Tip burn:** It is one of the most common physiological disorders found in Chinese cabbage which occurs due to the calcium deficiency by resulting rotting of leaf margins. The inhibited calcium absorption or translocation within the plant is the important factor for calcium deficiency as high salt concentration (especially high $NH_{4-}N$) or high moisture tension inhibit calcium absorption from soil which causes calcium deficiency in plant (Bose *et at.*, 2002). Calcium deficiency is mostly seen in young leaves with precipitation through rapid growth under low RH and high temperatures. It can be control with 0.7% calcium chloride + 50 ppm NAA by spraying 5 times (Wen *et al.,* 1991).

- **Blindness:** This type of plant is without terminal buds and leaves become large, thick, leathery and dark green colour. The main causes of blindness is due to low temperature during initial growing period and damage terminal bud at the time of transplanting as well as insect pest attack. Control measure is done by protecting the plant from low temperature injury and careful handling while transplanting (Choudhary *et al.*, 2009).
- **Bolting:** The phenomenon of premature seed stalk formation with head forming is known as bolting. There are lots of causes that can be responsible for bolting such as due sudden rise in temperature, varietal characteristics, failure of solid head formation, lack of pure seed etc. Control measure can be carried out by growing bolting resistant variety as well as avoid planting during warm season.
- **Loose head:** Loose head in Chinese cabbage is mostly occurs due to the temperature fluctuations during a specific development stage. The chances of this physiological disorder is maximum in the greenhouse when the temperature is high during the day time (above 25°C) and sometimes due to regular irrigations. So avoid planting during the warm season and give irrigation as per the requirement of plant and place.
- **Black speck.** It is also known as pepper spot, petiole spot and gomasho. This physiological disorder can be seen in both field and after the harvesting due to the low temperature (Masarirambi *et al.,* 2011). The symptom is occurrence of discolored lesions on both midrib and leaf vein. Pepper spot development can be reduced by storing Chinese cabbage with CO2 atmospheres (10%) (Cantwell and Suslow, 2001).
- **Chilling injury**. Chilling injury occurs when the Chinese cabbage is stored for more than 3 months at 0°C (32°F) (Cantwell and Suslow, 2001). The primary symptom, particularly on outer leaves, is midrib discoloration.
- **Physical injury**. It is due the poor handling during the packaging after the harvesting of the crop such as breaking or wounding of midribs causing browning and decaying process faster. Therefore, outmost care is needed during the handling and packaging before storing or transporting for the market purpose.

Plant Protection

There are attack of insect, pest and disease infestation is quite similar to other cole crops like cabbage, cauliflower, broccoli etc.

Harvesting and yield

Chinese cabbage is harvested 70-100 days after planning when the head of the plant become compact and erect even after pressing with hands (Jagdish, 2018). The harvesting should be done by cutting the fully develop head with knife from the base. Remove the outer loose leaves before transport for the marketing purpose. Chinese cabbage yields vary from 25-50 t/ha, depending upon the planting season, cultivar and site. The head of Chinese cabbage weigh 0.5-4.5 kg.

Post-harvest management

Post-harvest management of Chinese cabbage (also known as Napa cabbage) is an important aspect of the overall production process and can greatly impact the quality and shelf-life of the product. The following are some of the important practices for heading-wise post-harvest management of Chinese cabbage:

Cooling

Immediately after harvest, the heads should be placed in a cool place (ideally at a temperature of 32°F to 36°F) to slow down the rate of respiration and prevent spoilage.

Trimming

Remove the outer leaves and damaged or diseased portions of the heads before packaging.

Packaging

Place the trimmed heads in perforated polyethylene bags or other suitable containers to maintain a high humidity environment and prevent water loss.

Transportation

Transport the Chinese cabbage in a refrigerated vehicle to maintain the desired temperature and prevent spoilage.

Storage

Store the heads in a controlled atmosphere (CA) storage facility or a conventional refrigerator at a temperature of 32°F to 36°F and a relative humidity of 95% to 98%. This will help to maintain the quality of the product and extend its shelf-life.

Marketing

Market the Chinese cabbage as soon as possible after harvest to maximize its quality and market value. Consider labeling the product with its date of harvest and storage conditions to provide customers with important information about the product. By following these post-harvest management practices, you can help ensure that your Chinese cabbage is of the highest quality and has a long shelf-life.

Future prospect

The future prospects of Chinese cabbage look promising, especially given the growing interest in healthy and fresh food products. As consumers become more aware of the health benefits of leafy greens, the demand for Chinese cabbage is likely to increase. This versatile vegetable is rich in vitamins and minerals and is a good source of fiber and antioxidants, making it an attractive option for consumers who are looking to improve their diets. Additionally, with the rise of Asian cuisine and the popularity of dishes such as kimchi and stir-fries, the demand for Chinese cabbage is likely to continue to grow. Another factor that is likely to contribute to the growth of the Chinese cabbage market is the increasing popularity of home gardening and urban farming. As more people look for ways to grow their own food and become more self-sufficient, Chinese cabbage is a popular choice due to its ease of cultivation and high yield.

Overall, the future prospects of Chinese cabbage appear to be positive, with a growing demand for healthy and fresh food products, the rise of Asian cuisine, and the increasing popularity of home gardening and urban farming. These factors are likely to drive the growth of the Chinese cabbage market in the coming years.

References

Anonymous (2012) http: // ecoursesonline.iasri.res.in/mod/page/view php?id = 101559

Bose, T. K., Kabir, J., Maity, T. K., Parthasarathy, V. A. and Som, M. G. (2002). Vegetable Crops. Bhubani Mitra, Naya Prokash, 206, Bidhan Sarani, Calcutta. 1:345-392.

Cantwell M and Suslow T (2001). Cabbage Produce Facts. Postharvest Technology Center University of California, Davis.

Dhaka, R. S., Choudhary, B. R., Fageria, M. S. (2009). A Text Book on Production Technology of Vegetables. India: Kalyani Publishers.

Jagdish (2018). https://www.agrifarming.in/chinese-cabbage-farming

Kuo, C. G., and Toxopeus, H. (2016). *Brassica rapa* Chinese Cabbage (PROSEA). Plant Resources of South-East Asia.

Letard, M. (1984). Chinese cabbage. infos-CTIFL No.5: pp. 7-11.

Li, C. W. (1981). Chinese cabbage (Eds. T. S. Talekar and T. D. P. Griggs), Proc. First Intl. Symp. AVRDC, Taiwan, pp. 3-10.

Masarirambi MT, Oseni TO, Shongwe VD and Mhazo N (2011). Physiological disorders of Brassicas /Cole crops found in Swaziland: A review. African Journal of Plant Science; 5(1): 8-14

Wen, F. Y., Sun, D. L., Ju, P. H., Su, Y.M., An, Z. X. (1991). *Acta Horticulturea Sinica*, 18:148-152.

8

Advances in Production Technology of Onion

Pranava Praanjal[*1] and Mridulata Pant[2]

[1]*Ph.D. Research Scholar, Department of Vegetable and Spices Crops Uttar Banga Krishi Viswavidyalaya, Pundibari, Cooch Behar-736165, West Bengal*

[2]*Ph.D. Research Scholar, Department of Horticulture, Assam Agricultural University, Jorhat, Assam *Email: pranavapraanjal71@gmail.com*

Introduction

Onion is a cool season bulb vegetable of high economic importance. According to Vavilov, Central Asia is the primary centre of origin; maximum species were found in the Mediterranean region in the West to Central Asia in the East. In India, onion was introduced around 600BC and from here it was introduced to the rest of the world. The largest onion-producing states in India include Maharashtra, Madhya Pradesh, Karnataka, Rajasthan, Bihar, Gujarat, Andhra Pradesh, Haryana, West Bengal, and Uttar Pradesh, which together account for approximately 90% of the nation's total onion output. The expected share of Maharashtra in 2019-2020 is about 42.53% in onion production with 11363.00 tonnes. Onion has highest export value among vegetables, the total value earned in the year 2019 was ₹ 2660.79 crores (NHB, 2020). Onion contains about 89% water, 4% sugar, 1% protein, 2% fat. It also contains vitamin C, vitamin B6, folic acid and many other nutrients in trace amounts. Onion bulbs and onion greens are used in various dishes all over the world in cooked, as well as raw form. Red and yellow onions are used in soups because of their sweet taste. White onions are traditionally used in classic Mexican cuisine and also for dehydration purpose. India, Mexico, U.S., Turkey, Pakistan, etc are the major exporters of onion while the major importers of onion are Malaysia, UAE, Canada, Japan, etc.

Origin and Distribution

Origin of onion is Central Asia which comprises Iran, Afghanistan, Pakistan, Tadzhikistan and Uzbekistan. Mediterranean region is believed to be its secondary centre of origin.

Taxonomy

Onion is classified into the following categories:

Rank	**Scientific Taxonomical Details**
Kingdom	Angiosperms
Class	Asparagales
Family	Allioideae
Genus	Cepa
Kingdom	Angiosperms
Class	Asparagales
Family	Allioideae
Genus	*Allium*
Species	Cepa

BOTANY

Onion (*Allium cepa* L.) is a diploid crop species with chromosome number 16 (2n=2x=16). Although a natural tetraploid plant have been observed with 2n=32. Inflorescence of onion i.e. umbel may have 50 to 2000 flowers with average of 200-600 flowers. The length of flowering stalk is around 1-2 meters. A flower of onion contains 6 stamens, 3 carpels united into 1 pistil. The pistil has 3 locules and each locule has 2 ovules. Necteries is important characteristic of flower which contains nectar to attract pollinators i.e. insects for cross pollination. Protandry is found in onion and anther dehiscence occurs 3-4 days before the stigma become receptive which favours cross-pollination.

Several factors contribute to the characteristics and qualities of onions. One significant factor is pungency, which is attributed to a compound called allyl propyl sulphide. This compound is responsible for the sharp and strong flavor that is characteristic of onions. Additionally, onions can exhibit different colors depending on their variety. The red color in onions is primarily due to the presence of a pigment called anthocyanin, which gives them their vibrant hue. On the other hand, the yellow color in onions is attributed to the presence of a flavonoid called quercetin. This compound imparts a yellow tint to the onions and is also known for its antioxidant properties. In terms of their resistance to fungal diseases, onions contain an anti-fungal factor called catechol. This compound helps protect the onion bulbs from fungal pathogens, contributing to their overall health and longevity. The combination of these compounds and factors adds to the distinct characteristics and nutritional value of onions, making them a versatile and widely used vegetable in various culinary preparations.

Classification

The commonly cultivated onion species around the world are as follows:

1. **Common onion** (*Allium cepa* var. cepa) - with large bulbs born singly and propagated through seeds. It is mainly used as salad or in curries.
2. **Multiplier onion or potato onion** (*Allium cepa* var. *aggregatum*) - produces small bulbs borne in clusters and propagated by small bulbs. Mainly used for seasoning curries.
3. **Shallot** (*Allium cepa* var. *ascalonicum*) – This type of onion produces bulbs in clusters on the soil surface. Perennial in nature and usually do not produce seeds.
4. **Tree onion or Egyptian onion** (*Allium cepa* var. *viviparaum/ proliferum*) – The bulb-lets produced at the place of umbel and are used for seasoning and pickling. Bulb-lets are used for propagation.
5. **Chives** (*A. schoenoprasum*) (2n = 16, 24, 32) - This is a hardy perennial bunching herb grown for their hollow green leaves. It is propagated by root division.
6. **Welsh onion** *(A. fistulosum)* is another species which is very common in China and Japan and it is grown for its edible tops and leaf bases.

Soil and Climate

Soils such as sandy loam, silt loam and clay loam favour onion's growth. Mild alkaline soil with a pH range of 6.0-7.5 is preferred whereas saline, alkaline, highly acidic are not suitable. Proper drainage is very important for its good growth.

Extreme hot or cold as well as excessive rainfall is unfavourable for onion. The short-day onion is widely grown in plains, which requires 10 to 12 hours of light period while long-day onion requiring 13-14 hours of light period is grown in the hills. Temperature ranging from 13-24°C and 16-25°C is considered best for growth in vegetative phase and bulb development respectively.

Commercial Varieties

The production per unit area can be increased by growing high yielding varieties and adopting precision farming techniques. Two major group of onion varieties on the basis of day length and skin colour.

Red colour	**Yellow colour**	**White colour**	**Multiplier onion**
Pusa Red	Early Grano	Pusa White Flat	Co 1
Pusa Ratnar	Brown Spanish	Pusa White Round	Co 2
Pusa Madhavi	Arka Pitamber	Punjab 48 (S 48)	Co 3
Punjab Selection	Phule Swarna	Udaipur 102	Co 4
N53		Arka Swadista	MDU 1
Arka Niketan		Bhima Shweta	Agrifound Red
Arka Kalyan		Bhima Shubra	
Arka Bindu		Bhima Safed	
Agrifound Light Red			
Agrifound Dark Red			
Agrifound Rose			
Udaipur 101, 103			
Punjab Red Round			
Bhima Raj			

Varietal Characteristics of Some Important Cultivars

Arka Bindu	Bulb size 2.5-3.5 cm. Maturity- 100 days. Average Yield-250q/ha. Recommended for rabi season.
Arka Kalyan	Bulb size 4-6 cm. Maturity- 100-110 DAT. Average Yield-335q/ha. Recommended for kharif season.
Arka Niketan	Bulb size 4-6 cm. Maturity 145 DAT. Average Yield- 340q/ha. Recommended for rabi season.
Arka Pragati	Maturity-140-145 days. Average Yield 200q/ha. Recommended for rabi and kharif season.
Pusa Madhavi	Maturity 130-135 days. Medium to large bulbs. Average Yield 300-350q/ha. Recommended for rabi season.
Pusa Ratnar	Plant Height 30cm. Maturity 125 DAT. Average yield 400q/ha. Recommended for rabi season.
Pusa Red	Plant height 30cm. Bulbs medium-sized, flat to Zon globular. Maturity 140-145 days after transplanting. Average yield 250-300g/ha. Recommended for rabi season.

Pusa White Flat	Medium to large bulbs. Maturity 120-130 days. Average yield 300 q/ha. Recommended for rabi season. Suitable for dehydration.
Pusa White Round	Maturity 125-130 days. Average yield 300 q/ha. Recommended for rabi season. Suitable for dehydration.
Udaipur 101	Maturity 150-160 days. Average yield 200-300 q/ha. Recommended for rabi cultivation.
Udaipur 102	Bulbs 4.5-6.5cm in size. Maturity 120 DAT Rajas Average yield 300-350q/ha. Suitable for rabi season and dehydration.
VL 3	Plant height 50-70 cm. Average yield 250 q/ha Maturity 145 days. Recommended for rabi season.

There are several onion varieties and hybrids developed to meet different requirements and growing conditions. The Arka Kritiman and Arka Lalima are onion hybrids developed using the CMS (Cytoplasmic Male Sterility) system. These hybrids exhibit improved characteristics in terms of yield, quality, and adaptability. For thrips resistance, the N-53 and Pusa Ratnar varieties have been developed, which show resilience against thrips infestations, a common pest affecting onion crops. The Pusa Madhvi variety is developed through selfing and massing techniques, resulting in a stable and consistent onion variety.

For dehydration purposes, the Punjab 48 variety is preferred, known for its high Total Soluble Solids (TSS) content of 14.6%, making it suitable for the dehydration industry. Tri-parental synthetic varieties like Arka Bheem and Arka Akshay are developed by crossing three different parents, resulting in hybrid varieties with desirable traits such as improved yield and disease resistance.

For specific culinary purposes, there are recommended varieties such as Pusa Soumya, suitable for green onion production, and Arka Bindu, recommended for onion exports. Long day type varieties like Brown Spanish and Early Grano are suitable for regions with long daylight hours, promoting bulb development. Varieties like Arka Kalyan and Nasik Red exhibit resistance to purple blotch, a fungal disease common in onion cultivation.

Different seasons also have specific onion varieties, with Arka Kalian, N-53, and N-241 recommended for the kharif season. For both rabi and kharif seasons, the Arka Pragati variety is suitable, offering flexibility in cultivation schedules. These diverse onion varieties and hybrids cater to different agricultural and market requirements, providing farmers with options to optimize their onion production based on specific needs and environmental conditions.

Nursery, Land Preparation and Transplanting

The dimensions of onion nursery should be of length 3-4 meters and width 0.45 meters. Raised beds of height 15-22.5 meters should be made to avoid water stagnation. Soil treatment with Thiram @ 0.2% should be done and seeds should also be treated with Thiram @ 2-3 g/kg of seed just before sowing to avoid damping off disease. Line sowing is generally practiced. Spacing of 5-7 cm between the lines should be followed. Mulching with organic mulch like dry grass or straw should be done to maintain required temperature and moisture for germination. Nursery gets ready for transplanting in 40-45 days for kharif and 55-60 days for rabi season. Before dividing the field into beds, four to five ploughings should be done with certain intervals. Planking must be done to level the field. A width of 1.8 m is recommended for the beds. Seedlings can be transplanted in 8-10 weeks when they attain 15 cm height and 0.8 cm neck diameter.

Seed Rate

The recommended seed rate for onion grown in rabi season is 10-12 kg/ha while 12-15 kg/ha is recommended for kharif season.

Manures And Fertilizers

Application of 20-25 tones FYM at the time of ploughing in order to get equally distributed FYM all over the field. Onion requires high dose of nitrogen and potash. The general recommendation is 120 kg N, 50 kg P_2O_5, 50 kg K20 and 20 kg sulphur respectively. Nitrogen should be applied in 3 split doses i.e., at the time of transplanting, 30 and 45 days after transplanting.

Irrigation

The frequency of irrigation depends on the soil type and climatic conditions of an area but the general recommendation is up to 8-10 irrigations for kharif and 15-20 irrigations are given to rabi onions. Irrigation should be stopped 10-12 days before harvesting. Critical and most important stage in onion for irrigation is bulb formation.

Weed Management

One hand weeding is enough for weed management in the initial stages. Chemical management is necessary at later stages as onion is a poor competitor of weeds. Tenoran @ 2.5kg/ha after 3-5 weeks of transplanting and Basalin 1 litre/ha just after transplanting is effective for controlling weeds in onion.

Physiological Disorders and Their Management

Disorder	Management
Bolting Emergence of seed stalk prior to its time of formation generally due to extreme temperature fluctuations. Bulbs become light and fibrous. Keeping quality of bulbs is reduced.	• Grow non-bolting cultivars • The transplanting time should be such as to provide moderate temperature at the time of bulbing
Splitting and doubling of bulbs It is caused due to improper nutrient availability and adverse climatic condition. Mechanical injuries during inter cultural operations may also cause splitting of bulbs	To control splitting and doubling of bulbs in onions: • Select onion varieties with lower tendency for splitting and doubling. • Maintain consistent soil moisture levels through proper irrigation. • Provide balanced nutrition, avoiding excessive nitrogen during bulb formation. • Ensure adequate spacing between onion plants for proper bulb development. • Implement effective disease and pest management practices. • By following these practices, you can minimize the occurrence of splitting and doubling in onion bulbs.

Plant Protection

Major Insect-pests and Their Management

Pest	Symptoms	Management
Thrips (*Thrips tabaci*)	It can be identified by curling and twisting of leaves. Yield loss up to 50%	0.1% Malathion + 0.1% Sandovit (4 sprays at 15 days interval.
Onion maggot (*Hylemia antiqua*)	Infected plants turn yellowish brown and finally dry up.	Phorate 10G @ 20-25 kg/ ha should be applied before transplanting.

Major Diseases and Their Management

Disease	Symptoms	Management
Purple blotch (*Alternaria porri*)	Initially small, elliptical lesions or spots that often turn purplish-brown which are surrounded by chlorotic margin.	a) Dip the bulbs before sowing in 0.25% solution of DithaneM-45 for 1 hour. b. Spray Dithane M-45 @ 0.25% or Captan @ 0.2% at 10-15 days intervals from 30 days after transplanting or as soon as disease appears.
Onion smut (Urocystis cepulae)	Black lesions appear near base of scales and blisters near the base of leaves, downward bending of leaves occurs.	a) Treat the soil with fumigants before planting b) Seed treatment with fungicides like Captan or Thiram (2-3g/kg seed)
Yellow dwarf (Onion yellow dwarf virus)	The plant gets severely stunted. Leaves and stems lose green colour.	Malathion is recommended to control disease vector.
Onion smudge (*Colletotrichum circinans*)	Fungal disease mostly affecting white onion. Dark green to black spots of various sizes and shapes appears on outer scale of bulb.	Bulbs must be cured properly after harvesting and treated with 0.2 per cent suspension of Thiram, Captan or Difolaton.

Harvesting

The crop is ready for harvest in 5-6 months after transplanting depending on variety. In rabi onion, the bulb harvesting should be started when 50-75% neck fall occurs whereas in kharif season it is done based on visual inspection only. Harvesting is done by pulling out plants when tops are drooping but still green. Hand-hoe is used to pull out bulbs. During hot days when soil is hard.

Curing

Curing is a very important operation after harvesting in onion. To get rid of extra moisture from the outer skins, roots, and neck tissues of harvested onion bulbs, curing is done. It is a drying procedure. It enhances onion bulb preservation and lowers the risk of disease-causing organisms infecting the bulbs while they are being stored. It helps in increasing the storage life of onion by removing the excess moisture from bulbs. In north India, during the kharif season, when the temperature is low, the leaves and crop are both sun-cured for two to three weeks. If the temperature is very low, dryer is used and hot air blow for 16 hours at 46°C. A rabi crop is prepared by covering the bulbs with their leaves in the rows of the field for two to three days, trimming the leaves 2-2.5

cm above the neck, and then allowing the crop to cure in the shade for seven to ten days.

Yield

The yield of onion may vary among varieties and from season to season. Yield in kharif season onion is very low as compared to rabi onion. In rabi season the average yield obtained in between 30-35 t/ha while 20-25 t/ha is obtained during kharif season.

Storage

Onion bulbs are usually stored for a period of 4-6 months in India from May to November. Rotting losses occur maximum during the months of June-July. Proper grading, sorting and curing of the bulbs must be done before storage. Onion storage in ventilation condition is satisfactory at temperatures between 25-30°C and relative humidity range of 65-70%. Spray of Maleic Hydrazide (MH) is done to prevent sprouting.

Future Prospects

Significant scientific potential exists for future breakthroughs in onion growing. The industry covers a wide range of sectors for research and development that may support its development and sustainability. Genetic advancement, biotechnology, agronomic methods, pest and disease control, post-harvest preservation, and environmental impact reduction are important areas of scientific study.

The goal of genetic improvement through breeding programmes is to create onion varieties with desirable features including greater production, improved pest and disease resistance, improved stress tolerance, and prolonged shelf life. To speed up breeding and produce desirable genetic features, this requires using molecular markers, genomic selection, and gene editing approaches.

In order to produce genetically modified (GM) onions with improved properties, biotechnology is essential to onion agriculture. Benefits of GM onions may include better nutritional value and higher resilience to pests, diseases, and herbicides. However, to guarantee their safe deployment and customer acceptability, in-depth study, risk assessment, and adherence to regulatory frameworks are crucial.

To increase onion yields while reducing resource inputs and environmental effects, agronomic practises must be optimised. To minimise chemical inputs and improve sustainability, research focuses on effective irrigation techniques, precise nutrient management, crop rotation techniques, and integrated pest management tactics. Management of pests and diseases is essential to the production of onions. In order to create novel and environmentally friendly management techniques, scientific research is being done to better understand the biology, behaviour, and ecology of the diseases and pests that damage onions. The use of bio-control agents, resistant cultivars, cultural practises, and the selective use of non-toxic insecticides are examples of this.

In order to increase the shelf life of onions, lower losses, and preserve their quality throughout storage and transit, post-harvest preservation procedures are used. The application of post-harvest treatments such modified atmospheric packing, edible coatings, and natural antibacterial agents are the main topics of scientific research on optimised curing, storage conditions, temperature control, and humidity management.

A major scientific problem is addressing how onion farming affects the environment. Utilising effective irrigation systems, reducing greenhouse gas emissions through the adoption of sustainable practises, enhancing fertiliser application to lessen nutrient runoff and water pollution, and creating environmentally friendly alternatives for pest and disease management are all goals of the research.

References

Anonymous. Onion (2003). Accessed from http: //www.nhb.gov.in.

A. Farha. B. Shahanaz and AFM Jamal Uddin (2020). An overview of studies on summer onion production technology. *International Journal of Business, Social and Scientific Research.* *8*(**3**): 25–27.

Bahnasawy A H. 2016. Onion losses during storage as Influenced by curing method. https:// adelbahnasawy.files.

Bose, T. K., Kabir, J., Maity, T. K., Parthasarathy, V. A., Som, M. G. (2003) Vegetable Crops. 3rd, 3, Naya Udyog, Kolkata, 3, 5-118.

Choudhary, Desh Raj. (2018) Scientific Cultivation of Onion (*Allium cepa L.*). Brillion Publishing, New Delhi, 239-260.

Choudhary, Mahesh, B. L. Asiwal, and R. K. Dular. "Knowledge level of farmers about improved production technology of onion crops in sikar district of Rajasthan." *Journal of Krishi Vigyan*. 8.1 (2019): 191-196.

Chadha, K. L. (2019). "Handbook of horticulture", Volume 1, Indian Council of Agricultural Research, New Delhi, 477-582.

Kamble, Ankush, and Utkarsh Tiwari. (2019). "Instability in Onion production and marketing in Maharashtra." *International Journal of Science and Research*. 8(2): 975-979.

Muthukumar, P., Selvakumar, R. (2017) Glaustas Horticulture. 2nd, New Vishal Publications, New Delhi, 297-303.

Padula, Gregorio, Xianzong Xia, and Roman Hołubowicz. "Welsh onion (*Allium fistulosum* L.) Seed physiology, breeding, production and trade." *Plants*. 11.3 (2022): 343.

Peter, K. V. (1998) Genetics and Breeding of Vegetables. 1st, Indian Council of Agricultural Research, New Delhi, 13-20.

Ram, H. H. (2012) Vegetable breeding: principles and practices.*3rd*, Kalyani Publishers, Ludhiana, 539-569.

Singh, D. (2019) Vegetable Science. 1st, New Vishal Publications, New Delhi, 129-143.

Thamburaj, S and Singh, N. (2015) Textbook of Vegetables, Tuber and Spices. Indian Council of Agricultural Research, New Delhi, 165-187.

Thompson, A. K., Booth, R. H and Proctor, F. J. 1972. Onion storage in the tropics. Tropical Science 14: 19-34.

Walker J. C. (1921). Onion Smudge. *Journal of Agricultural Research*, 20(9).

9

Advances in Production Technology of Garlic

Dibya Jyoti Mukhia[1*], Ranjit Chatterjee[2], Suprava Biswal[3] and Raj Kumar[4]

[1,3&4] *Ph.D. Research Scholar, Department of Vegetable and Spice Crops, Uttar Banga Krishi Viswavidyalaya, Pundibari, Cooch Behar-736165, West Bengal*

[2]*Professor, Department of Vegetable and Spice Crops, Uttar Banga Krishi Viswavidyalaya, Pundibari, Cooch Behar-736165, West Bengal*

[*] *E-mail: divyamukhia8@gmail.com*

Introduction

Garlic (*Allium sativum* L.) belongs to the monocotyledonous family Alliaceae or Amaryllidacea; Chromosome No. 2n=2x=16. In India, it is the second most extensively grown bulb crop after onion. According to several botanists (Candolle, 1985; Dscalov and Popov,1941; Kuznetsov, 1948 and Regel, 1975) garlic was cultivated and utilised by Mediterranean population since medieval time.

Origin, History and Distribution

Central Asia, particularly elevated highland of Pamir and Tian Shan is considered as the primary centre of origin of garlic (Vavilov, 1926), however Caucasus and North Africa is considered as the secondary centre of origin. *Allium longicuspis* is considered as the wild ancestor of garlic. The history of garlic usage dates back to 3000 B.C when it was used in Egyptians tombs likewise its first usage in modern era occurred during 10th century in England (Shinde and Sontakke, 1986) and its first introduction in North and South America was done by Spanish and Portugese settlers during fifteenth century; introduction and usage of garlic in other European and Asian countries occurred subsequently.

Area and Distribution

In 2016, the world production of garlic bulbs was reported to be around 27 million tonnes, where China contributed nearly 80% of the global production to the extent of 21 million tonnes (FAOSTAT, 2016). The world production of garlic bulbs was reported to be around 12 million tonnes and the major share to the extent of 10 million tonnes being that of Asia (Anon., 1998). The important garlic producing countries in the world are China, India, Bangladesh, Egypt, South Korea, Russia, etc. The important garlic producing countries in the world are China, India, USA, Russia, Turkey, Spain, Thailand, Korea Republic, Egypt, etc. China ranks first in garlic area and production followed by India and Korea Republic. In India, Rajasthan is the leading state in terms of production share of garlic (52.87%) followed by Uttar Pradesh (17.13%), Gujarat (13.72), Punjab and Assam (NHB, 2019). In India, major garlic growing states are Gujarat, Orissa, Madhya Pradesh, Rajasthan, Uttar Pradesh and Maharashtra.

Nutritional Composition and Usage

Among the several bulb crops garlic has considerably high amount of nutrients such as carbohydrates, protein, phosphorus, ascorbic acids (listed below in Table 1) and water soluble amino acid like "allin", which has several anti-bacterial, anti-thrombic and health promoting properties. Allin is colorless and odourless water-soluble amino acid, on the injury of garlic cells the enzyme allinase gets activated and breaks down stable allin to allicin which is an unstable, sulphur containing compound (diallyl disulfide) which imparts garlic its typical odour (Nasir *et al.*, 2020). Composition of fresh peeled garlic cloves (bulblets) and dehydrated garlic powder as reported by Pruthi (1979) is given in Table 1. The vitamin C content in garlic varies with cultivars, mean content being 315 mg/kg fresh matter (F.M.), cv. Vladan showed the lowest amount (273 mg/kg F.M.), while the highest one was found in cv. Vekan (403 mg/kg F.M.) (Pokluda and Petrikova, 2001). Shankaracharya (1974) reported that garlic contained about 0.1% volatile oil. The chief constituents of the oil are: diallyl disulfide (60%),diallyl trisulfide (20%), allylpropyl disulfide (16%), a small quantity of diethyl disulfide and probably diallyl polysulfide. Diallyl disulfide is said to possess the true garlic odour. Garlic is also a rich source of several plant secondary metabolites particularly quercetin, kaempferol, myricetin, luteolin which are potential antioxidants (Tahir *et al.*, 2022).

Table 1: Composition of garlic*

Particulars	Fresh peeled garlic cloves (bulblets)	Dehydrated garlic powder
Moisture %	62.80	5.20
Protein %	6.30	17.50
Fat %	0.10	0.60
Mineral matter %	1.00	3.20
Fibre %	0.80	1.90
Carbohydrates %	29.00	71.40
Calcium %	0.03	0.10
Phosphorus %	0.31	0.42
Potassium %	-	1.10
Iron %	0.001	0.004
Niacin %	-	0.70
Sodium %	-	0.01
Vitamin A (I.U.)	0.0	175.00
Nicotinic acid (mg/100 g)	0.40	-
Vitamin C (mg/100 g)	13.00	12.00
Vitamin B (mg/100 g)	-	0.68
Vitamin B_2 (mg/100 g)	-	0.08

*Pruthi (1979).

Importance and Uses

Worldwide, Indian cuisine has wider acceptability because of its rich flavour, aroma and pungency; garlic is the integral part of the Indian cuisine and has been traditionally used as spice or condiment in preparations of many foods, most notably chutney, pickles, vegetable curry, seasoning dal, meat and fish preparation. In Philippines, much of Eastern Asia and in parts of tropics, the green tops as well as bulbs of garlic are used. Due to change in food habits, adoption of westernized diet and to account postharvest losses (nearly 40% in terms of perishable horticulture products) dehydrated garlic in powdered or granulated form is being extensively used in several food processing industries for tomato ketchup preparation and for flavouring several value-added products like ramen, noodles, sausages and sauce etc.

Since ancient times garlic was utilised by Egyptians as an ingredient for both culinary and medicinal purposes. In India, the *Unani* and *Ayurvedic* medicinal system is widely practiced and they consider garlic as carminative and gastric stimulant which aids in digestion and absorption of food. Garlic is characterized as having several anti-bacterial, anti-fungal, nematocidal and several insecticidal properties. Apart from industrial its usage in several food preparation, garlic oil also has medicinal importance as it helps in treating pulmonary tuberculosis, rheumatism, sterility, impotency, cough, red eye and controlling blood sugar. Fresh garlic on steam distillation yields 0.1-0.5% essential oils.

Botany

Garlic is a frost tolerant, perennial bulbous herb that stands upright and is 30-100 cm tall with narrow flat leaves which bears small white flowers and bulbils (Janick,1979).It is a herbaceous annual that produces bulb and biennial seeds. In contrast to onion, which has hollow scape, garlic scapes are smooth, spherical and solid over their whole length. Cloves are the 6-35 smaller bulblets that make up the bulb, which is encased in a thin layer of whitish or pinkish pappery sheath. Most of the Garlic cloves does not grow blossom stalks. Globose, frequently with bulbils, many flowers, long peduncle; umbel. Oblong, acuminate, greenish white or slightly tinged with purple sepals are present.

Taxonomy

The botanical name *Allium sativum* was given to garlic by Carl Linnaeus in 1754. It is classified under class monocotyledon and belongs to the section *Porrum*. Garlic is classified as bolting and non-bolting type. Under this classification the non bolting species is *Allium sativum* var. *sativum* and *Allium sativum* var. *vulgare* and the bolting species are *Allium sativum* var. *ophioscorodon* and *Allium sativum* var. *scordoprasum*. It was only in 1948 Kuznetsov gave botanical and ecological classification of the species;the two sub-species was *Allium sativum* var. *sagiatum* Kuz. (Bolting garlic) and *A. sativum* var *vulgarem*Kuz. (non-bolting garlic). According to Nair *et al.* (2013) garlic species are classified into four groups: *A. longicuspis*, *A. ophioscorodon*, *A. sativum*, *A. subtropical* and *A.pekinense* sub-group. The *longicuspis* group is considered the oldest and progenitor of garlic. The *A. ophioscorodon* group is distributed in Central Asia, the *sativum* group in the Mediterranean zone and the *subtropical* in the south and southeast of Asia. The *pekinense* group comes from the east of Asia. Most of the cultivars grown in India are short day type with critical day length requirement of 12hours; long day type are grown at higher elevation. Generally, garlic bulbs are creamish white in color but pink types are also found; commercial varieties bear multiple cloves of 3.0-3.5cm diameter and cloves ranging from 20-40, however some of the local cultivars bear single cloved bulb too.

The elephant garlic or great headed garlic (*Allium ampeloprasum;* hexaploid 2n=2x=48) is not a true garlic, but a type of leek that produces larger bulb with fewer larger cloves (4-8). Several small cormels may also develop on the bulb near root zone (5-10). Plants have flat and thin leaves and produced a large scape and umbel (flowering head) but no seeds. It gives milder flavor which is intermediate between garlic and onion. Allicin content of great headed garlic is less than common garlic. The cloves are largely used for medicinal purpose.

a. **Bolting Subspecies:** This sub-species of garlic is commonly found in all parts of the world. The sub-species is further classified into three ecotypes:
    1. **Asian Ecotype:** It is referred as 'multi-flowered' ecotype with the pseudo stem of about 150cm height or more. A vast number of sterile blooms and hundreds of additional little aerial bulbs make up the inflorescence.
    2. **Caucasian Ecotype:** The flower stem is of average height of 60-80cm and bears large inflorescence consisting of a small number of sterile flowers and pea sized aerial bulbs. According to the various research the bulbils planted in autumn form bulbs during the next year.
    3. **Seaside Ecotype:** This ecotype is distinguished by delayed formation of flower stem, which usually appears after formation of three of four fertile leaves at the base of which two or three circles of cloves are formed. Compared to the other two eco-type the vegetative period is prolonged by 12-15days due to delayed formation of flower stem.

b. **Non-bolting Subspecies:** This type of garlic doesn't bear flower and is also classified into three ecotypes:
    1. Continental Ecotype: It is characterized by its longest vegetative period, thin and compact pseudo stem of about 40cm height, thick, waxy and smooth leaves, bulbs differ in the size and number of sterile leaves (8-10)
    2. South Russian Ecotype: It is distinguished by its good storability and early maturity; the plant is xerophytic type with thin, tall (upto 45cm) pseudo stem
    3. Early Mediterranean Ecotype: This cultivar is named so based on the area in which it is cultivated. The unique feature of this ecotype is its mesotype structure which is greatly influenced by the surrounding ecological factors, moisture particularly.

Soil and Climate

Garlic being a bulbous crop prefers medium black, fertile, well drained loamy soil rich in humus with good potash content. The bulbs grown in sandy or loose soil are lighter in weight whereas if the crop is grown under heavy clay soil it leads to the production of misshapen bulbs and bruised or broken bulbs due to difficulty in harvesting The neutral pH soil (6-7) is highly suitable for the crop; the performance of bulb is not good in alkaline and saline soil.

Garlic is a crop capable of growing in diverse climatic conditions but grows best at the elevation of 1000-1300m above mean sea level. In India commonly short day type varieties are cultivated. Even though it is a frost hardy and a cool season crop it is unable to stand extreme cold weather conditions and prefers cool and moist period during growth and relatively dry period during maturity of bulb. On exposure of bulbs to low temperature it leads to development of cloves in the leaf axil near the bulbs surface i.e further from bulb centre than usual. The temperature requirement of the bulb varies according to the growth stage of the crop. In the initial vegetative stages, it prefers comparatively lower average temperature i.e 10-15°C as compared to the bulb initiation stage i.e 25-30°C. The critical day length for bulbing is 12 hours. Long days with high temperature favours bulb initiation whereas short days leads to development of secondary bulbs and suppressed bulb formation. To break the dormancy of cloves or planting material the exposure of them at aroung 20 C for 1-2 months hastens bulbing.

Field Preparation

The field is to be prepared by ploughing 4-5times repeatedly. Being a shallow rooted crop, the roots don't penetrate to the depth of 8-10cm so shallow ploughing is preferred. Planking is to be done for proper levelling. For the ease in irrigation activity the field is divided into small plots and channels. The width of the bed should be 1.8m and length is to be made according to the convenience in the intercultural operations and according to the level of the land. The plant spacing is maintained according to the variety. A spacing of 10cm X 8 cm is suitable for small sized varieties and 15 X 10cm spacing is desirable for varieties with large bulbs. Too wide spacing results in thick neck bulbs, whereas the narrow spacing favours disease attack like purple blotch.

Sowing Time

India being a country of rich agro-climatic diversity the sowing time of the crop varies from region to region. But as the crop cannot tolerate excess heat or cold it is condusive to sow the crop in winter and harvest before the on-set of summer in plains.

Regions	Sowing Time	References
North India (Plains)	September-October	Rao and Purewal, 1957
North India (Hills)	March-April	Rao and Purewal, 1957
South India	August-November	Rao and Purewal, 1957
Gujarat, Assam, West Bengal and some parts of Orrisa	October-November	
Maharastra	November-December	Mathur *et al.*, 1975
North Bihar	October-December	Singh *et al.*, 1984
Northern district of West Bengal	2nd week of November	Das *et al.*, 1985b

Methods of Planting and Spacing

Garlic is vegetatively propagated by detached cloves of size 8 to10mm, bulbils and tissue culture particularly meristem tip culture are used to produce healthy seedlings. If the bulbs are planted then the bigger sized cloves on the outside of the bulbs are selected for planting, long slender cloves to be discarded as they give poor yield. For one hectare land about 300-500 kg of cloves is required; the quantities is variable depending on the individual clove size and planting distance. In general garlic is chiefly propagated by five methods viz., (a) Dibbling method (b) Furrow planting (c) Broadcasting (d) Transplanting (e) Flat vs ridge planting.

a) Dibbling: This method is widely practiced for cloves of small to medium areas. Placing the sprouted end upwards the cloves are dibbled at the depth of 5 to 7.5 cm deep; the recommended spacing is variable 10 x 15 cm, 10 x 20 cm, 15 x 18 cm. The light irrigation is required immediately after sowing.

b) Furrow planting: Using hand hoe or cotton drill the furrows are made 15 cm apart; individual cloves are planted in these furrows at 7.5 to 10 cm apart followed by covering with soil and light irrigation. Gives poor yield.

c) Broadcasting: The land is divided into well levelled seed beds and the cloves are broadcasted, after that it is covered by harrowing followed by light irrigation immediately after sowing. After 3-4 days of sowing the uncovered cloves are covered in the soil followed by another irrigation. Doesn't give good yield

d) Transplanting: According to Maurya (1987) the seedlings of 30days old when transplanted on 15th November gave the same yield as the directly sown cloves. But delay on transplanting beyond 22nd of November resulted in reduction of bulb yield.

e) Falt Vs. Ridge planting: In this method the ridges are prepared at 35 cm spacing and the cloves are planted at both the sides at 7.5 cm distance; plant spacing of 15 x 7.5 cm is maintained in flat beds.

Commercial Varieties

The variety of garlic are classified based on its color i.e red and white; amongst the two the white colored types are popular as they are less pungent whereas the pink colored ones are widely used for the medicinal purpose.

Among the white colored variants they are again classified based on its size i.e small sized bulb and big sized.

A. Small sized varieties: T 56-4, Creole, Italian and Tabiti.

B. Big sized varieties: Fawari, Rajalle Gaddi, Solan, Madrasi, Jamnagar Local, Ladwa, Aru, Sweta etc.

Characteristics of few of the varieties are listed below:

Creole: It is an early variety compared to Italian with broad leaf and green top, shorter stem, less wrapper leaves and large cloves. It is a high yielder with inferior keepimg quality.

Fawari: It is a white colored, big sized, compact bulb bearing, high yielding variety. It is popular in south India particulary Bellary region of Karnataka.

Rajalle Gaddi: It is a commonly cultivated variety of Karnataka and Andra Pradesh. The bulbs are medium sized and pungent in taste with good keeping quality.

G-6: This variety is tolerant against soil salinity.

Godavari: This variety gives good yield but the physiological disorder of premature sprouting is prominent. Produces pink colored bulb.

Madrasi: It is a local variety of Rajkot region of Gujarat.

Agrifound Parvati (G-313): This variety was created through the judicious selection of a rare assortment of Hong Kong market. This variant is a long day kind and therefore it can be grown in mid and high hill of northern states. Bulbs are bigger sized (5-6.5 cm), the diameter of clove is 1.5cm to 1.8cm with 10-16 cloves found in one bulb, creamy white with pinkish tinge. Suitable for cultivation in hills. Average yield of the variety is 17-255 q/ha.

Agrifound White (G-41): Local accession was collected from the Sharif region of Bihar and were used to develop this variety at NHRDF, Nasik. Bulbs have cream coloured flesh and are silvery white with a bulb diameter of 3.5 to 4.5 with the size index of 12-15cm. The cloves of this variety are enormous and elongated with 20-25 cloves/bulb. This variety has 42.78% dry matter, 41%

TSS and excellent keeping quality. Yield of the variety is 130q/ha. Tolerant to *Stemphylium* disease and purple blotch.

Fawri: Being a high yielding variety it is a popular in south India and West Bengal.

Jahiti: This variety bears darkest and largest cloves.

Ooty -1: Bulbs matures in 125-130days, are dull white in color, bigger in size, bears 20-25 cloves/bulb, is resistant to thrips and yield is 17 t/ha.

Pant Lohit: This variety is moderately resistant to purple blotch and the cloves have distinguishable pink tinge color.

T 56-41: The bulbs are white in color and medium sized.

G-282: Local accessions was collected from Dindigul (Tamil Nadu) and was used to develop this variety at NHRDF, Nasik. This variety is believed to perform well in northern and central part of India. In comparison to the other variety this variety has wider leaves, with creamish white bulb of 5-6cm diameter consisting of15-16 cloves/bulb. This variety has medium storability with 39-43% dry matter content and TSS OF 38-41%. Average potential yield of the variety is 175-200 q/ha.

G-50: The performance is good in northern and western India. Maturity: 165-167 days. Tolerant to *Stemphylium* disease and purple blotch

Jamnagar: This variety is widely cultivated in Gujarat, Maharastra, Andra Pradesh and Madhya Pradesh. This variety has bigger sized clove.

Rajalle Gaddi: In comparison to the other varieties grown in South India this variety is a bit bigger.

T-56-4: This variety is suitable for Punjab and Uttar Pradesh.

Yamuna Safed (G-1): Local accessions was collected from Delhi (Azadpur) and was used to develop this variety at NHRDF, Nasik. Bulbs are sickle shaped, have creamish flesh with silvery white skin; it has bulb diameter of 4.0-4.5 with the size index of 12-15cm. This variety is tolerant to insect like thrips and diseases like purple blotch and has a yield potential of 150-175q/ha.

Yamuna Safed-2 (G-50): It is developed from NHRDF, Nasik by the collection of local accession obtained from Karnal area of Haryana. This variety is mostly recommended for the northern areas.

Punjab Garlic: Cloves are white in color and leaves dark green. Bulbs are uniform and attractive and gives a yield of 90-100 q/ha.

Garlic 56-4: Characteristic feature is the central stalk bears 3-5 dark red shinned bulbils at the top. Cloves are 25-34 in number. Yields 80-100 g/ha.

Varieties developed from different institutions or organizations are given below:

MPKV, Rahuri: Godavari, Sweta (both are short day type)

CCSHAU, Hissar: HG-1 and HG 6.

IARI, New Delhi: Pusa Sel. 10/

PAU,Ludhiana: LCC1

VPKAS, Almora: Aru 52

NHRDF: Agrifound white (G-41), Yamuna Safed (G1), Yamuna Safed 2 (G 50), G 282 and Agrifound Parvati.

DOG, Rajgurunagar: Bhima Omkar, Bhima Purple

TNAU: Ooty 1 and Ooty 2

Acclimatise variety: Early White or Medican Varieties, Creole and Tatule and Italian.

Local type: Singapore red, Rajali, Madrasii, Mettupalayam

Varieties suitable for export: Agrifound Parvati, Yamuna Safed, Agrifound White, Jamnagar Ladwa, Malek.

Varietal Improvement

Garlic is a vegetatively propagated crop and there isn't much scope for its improvement naturally, clonal selection using local types is one of the methods by which local U.P cv. Chaubattia-3 was proven to be promising clonal selection in terms of bulb size and yield. In different regions of India excellent local strains are available I large numbers and they vary in number of cloves; ranging from 16-50 and the size of bulbs (Nath, 1976). Various breeding methods, in-vitro polyploidy, biotechnological tools have been employed for developing variation and crop improvement. Especially in case of several viral resistance

a. **Breeding:** It includes novelization of varieties, screening of varieties and development of hybrids. Most the garlic breeding work in India first started with screening of the local variants for its performance, biotic and abiotic stress resistant abilities.

- Mishra *et al.*, 2013 evaluated twenty promising garlic accessions for growth, yield, quality and resistance to *Stemphylium* blight disease under Northern Indian condition. They found that the lines G-189, G-176, G-302, G- 304, G-369, G-366, G-222 and G- 378 were found promising compared to check varieties (G-1, G- 41, G-50, G-323 and G-282).

- Prajapati *et al.*, 2018 evaluated thirty diverse garlic genotypes for morphological characters. The genotype JAS 28 recorded the highest value for most of the quantitative traits such as, leaf length (36.33cm), leaf width (1.74 cm), pseudostem diameter (0.84 cm), average weight of bulb (27.0 g), marketable yield (117.0 q/ha) and total yield (118.0 t/ha). However the maximum plant height (62.20cm), number of cloves/bulb (23.66) and average weight of bulb (27.0 g) were recorded by the genotype JAS 16.

b. Polyploidy *in vitro*:

- Jang *et al.* (2000) made attempts to induce polyploids by colchicine treatment in clones of Namdo-garlic. Calli. Results showed that the optimum concentration and period of colchicine treatment for inducing polyploidy was 1.0 mM for 6-10 days. After acclimatization, initial growth of polyploid was slower than that of diploid but its growth speed recovered in the late stage.

c. Biotechnology:

Micropropagation and *in vitro* bulb formation

- Kehr and Schaeffer (1976) could successfully produce differentiated plantlets by tissue culture from growing points of bulbs.
- Xu*et al.* (2000) investigated the various factors affecting the efficiency of rapid multiplication of virus-free garlic by inflorescence meristem culture and multi-bulbil induction. A total of 5000 multi-shoots or multi-bulbils were obtained from one pedicel per year through 4 cycles of inflorescence meristem culture. Garlic genotype and culture medium significantly affected the number of multi-shoots, while the day length affected multi-bulbil induction.
- Xiong *et al.* (2000) presented the in vitro propagation of virus-free *A. sativum* plantlets. Suitable subculture conditions included MB medium supplemented with BA at 2 mg/l, NAA at 0.1 mg/l, gibberellic acid at 0.05 mg/l, and sucrose at 20 g/l at pH 6.2.
- Resende *et al.* (1998) reported that five garlic cloves, numbered 8, 9, 14, 15 and 28, of cv. Gigante Roxo derived from tissue culture exhibited degeneration after 4 generations of standard propagation in the field

Somatic embryogenesis and somaclonal variation

- Ohsumi *et al.*, 1993. Interspecific hybrids between *A. cepa* and *A. sativum* were obtained using the fertile clone *A. sativum* as the male parent. The nascent embryos which formed shortly in interspecific hybridization between *A. cepa* and *A. sativum* were rescued by ovule culture at an early stage. The zygotes or proembryos developed in MS medium. The hybrids produced not only S-propenyl-L-cysteine sulfoxide, which is the major flavour precursor in A. cepa, but also S-allyl-L-cysteine sulfoxide (alliin), which is characteristic of *A. sativum*.

Biochemical and molecular markers

In the contemporary research on the genetic diversity of garlic, molecular markers such as RAPD (Restriction fragment length polymorphism), AFLP (Amplified fragment length polymorphism),SSR (Simple sequence repeat), SRAP (Sequence related amplified polymorphism) and ISSR (Inter-simple sequence repeat) are largely uses to assess the diversity and relationship among the different garlic genotypes as the molecular markers are extremely sensitive and neutral to environment effects and crop management practices.

- Etoh *et al*., 2001 compared RAPD variation of 30 garlic cloves from the primary centre of origin (Central Asia) was compared with that of 30 garlic cloves from the western most area of distribution (Iberian Peninsula).
- (Bradley *et al*., 2001). Evaluation of genetic similarities in 20 Australian garlic cultivars was studied. Using DNA fingerprinting, the cultivars were differentiated into bolting, non-bolting or intermediate bolting types and into early-, mid- and late-season types. The range of genetic similarity between the 20 garlic cultivars varied from 58 to 97%, confirming a high degree of genetic variation.

Nutrition Management

Application of well decomposed FYM @ 300 q/ha during field preparations, 100-200:50:50-120 kg/ha NPK. Where full dose of P and split dose of N & K at the time of planting and another 50% again subdivided into two split each after 60 days after sowing and 90 days after sowing as top dressing. Soil application with borax upto 10kg/ha increases bulb size and yield. Foliar application with micronutrients like manganese (0.1% $MnSO_4$), boron (0.2% boroc acid), copper (0.02% $CuSO_4$) and zinc ($zNso_4$) help in dry matter accumulation in cloves.

Use of Plant Growth Regulators (PGRs): Application of PGR in garlic has proven to improve the yield and is listed below.

Name of PGRs	Concentrations (mg/l)	Method of application	Attributes Affected	References
Ethephon	500	Foliar Application at 20-25 DAS	Increases clove size and yield	Tiwari *et al.*, 1984
Alar	500	Foliar Application at 20-25 DAS		
CCC	1000	24 hours after planting	Increases yield and storability of bulb	Das *et al.*, 1985 b
NAA	50	60 and 90 DAS	Increases bulb yield	
MH	2500	Foliar spray at fortnight before harvesting	Under storage sprouting is checked	Om and Awasthi 1977
GA_3	200-400	Foliar spray	Stimulated formation of lateral buds	Tayagi and Aoba, 1976

Irrigation

Being a shallow rooted crop, water logged condition is unfavorable for growth and development of garlic. Immediately after planting the first irrigation is to be given followed by irrigation at weekly interval during vegetative stage and at 10-15 days interval until bulb development. Bulb development stage is the critical stage of water requirement. If the irrigation is given after a long dry spell, it will cause initiation of tops from partially developed cloves. Final irrigation is to be given three days prior harvesting as it aids in uprooting the bulbs without any damage.

Weed Management

Proper weed management is important to obtain considerable high yield as weed causes 40% or more losses. First hand weeding is to be done 30days after sowing followed by weeding at every monthly interval. Hoeing is the intercultural operation which has proven to be beneficial for the crop at 2.5months after sowing as it facilitates in loosening of soil for bigger and compact bulbs setting. However, hoeing in later stages can have a detrimental effect on quality of bulbs. Several weedicides such as Tribunal @ 1.5kh a.i/ ha, nitrofen @ 2.0 kg a.i/ha or basalin @ 2-2.5 kg a.i/ ha diluted in 625 liter can also be procured to control weeds after 7-15days of planting.

Physiological Disorder and Their Management

In case of garlic there are two physiological disorder i.e rubbering of garlic before harvest and premature sprouting of garlic bulbs.

A. **Rubbering:** It is a physiological disorder occurring at higher incidence in Tamil Nadu hills. It occurs in fresh bulbs at harvest, characterized by scrape development on applying pressure with tip of the thumb. On release of the pressure the scrape returns to original form; on drying the bulbs becomes spongy.

- **Cause:**

a) Excessive application of nitrogen in the form of urea.

b) Common in lowland, watershed area rich in deposition of nutrients and moisture caused by heavy rain.

c) Soil with excess moisture content caused by frequent irrigation than recommended level.

d) Short duration type (Mettupalayam type) is more susceptible than long duration type (Singapore type)

- **Control:**
- It can be controlled by judicious application of nitrogen fertilizer and its application in the form of ammonium sulphate instead of urea
- Application of nitrogen and potassium in split dosage @ 50% as basal dose and rest 50% in two splits of 45 and 60 days after planting
- At 30th day of planting spray of 1500 ppm (1.5g/ 1litre of water) CCC or MH will reduce the incidence.
- At 30th, 60th and 90th days after planting application of 0.3% Boron, 0.1% Sodium molybdate.

B. **Premature sprouting:** After the initiation of cloves instead of bulbing the cloves begin to develop leaves and prematurely sprout in the field.

- **Cause:**

Delayed harvesting during rainy season causes splitting and premature sprouting of bulbs.

High soil moisture during maturity caused by heavy rain or excess irrigation.

Wider spacing leads to excess water and nitrogen uptake by individual plants causing premature sprouting.

- **Control:**

a) Maintaining closer spacing of 15 cm x 7 cm during cloves planting.

b) Avoid excessive irrigation.

c) After 30 days of planting application of burnt ash @ 50 kg/ha.

Major Insect and Pest and Their Management

Aphid (*Myzus persicae*): Aphids are detrimental to garlic. It can be controlled by spraying Malathion @ 0.1%.

Mites (*Rhizoglyphus echinopus*): It is a mostly transmitted in seed bulbs and have been proven to be detrimental for growth and yield of garlic. It can be controlled by spraying of hostathion @ 0.05% or nemafos @ 0.15%.

Nematodes (*Ditylenchus dipsaci*): By soil application of 46 each at 30kg/ha of temik and nemaphos before or during planting the nematode population can be controlled. Pre-treatment of cloves with nemaphos-46 @ 0.15% helps in reduction.

Thrips (*Thrips tabaci*): Methyl dematon 25 EC 1 ml/lt

Major Disease and Management

Purple blotch (*Alternaria porri*): Above ground parts of plant are affected causing poor development of bulbs. Firstly, it appears in oldest leaves with the occurance of tiny, elongated, deep and whitish spot with purplish centre. Under favorable conditions these spots enlarge, coalesce and covered with black fruiting body of fungus.

Control:

- Seed treatment with Bavistin/ Thiaram @ 2.5g/kg followed by crop rotation.
- Good drainage condition should be provided during ntire crop growth period.
- Summer ploughing with reduced disease severity.
- Fungicidal application with Dithane M-45 @ 0.25% (Gupta *et al*., 1987)

Stemphylium leaf blight (*Stemphylium vesicarium*): It is a foliar disease in which the infection occurs at middle of the leaves which is small yellow to pale orange in color which later on turns brown. Leaf dries prematurely affecting bulb development.

Control: Application of Dithane M-45 @ 0.25% (Gupta and Pandey, 1986)

Clove rot (*Fusarium oxyxporum* f. sp. Cepea): pretreatment of cloves with Carbendazin @ 2g/kg.

Yield

It is variety specific and depends largely on nutritional management, irrigation, climatic parameters and insect and pest management. But it varies from 50 to 100 q/ha.

Harvesting and Postharvest Management

The harvesting stage of the garlic is when the tops turn brownish or yellowish, showing signs of drying up and bending. It takes nearly 130-180 days after planting depending upon the variety, nutritional management and season. Early harvesting results in poor quality bulbs where loss is caused by drainage. Delayed harvesting results in spliting and resprouting of bulbs. In Indian condition the harvesting is done manually using hand digger. Bulbs are harvested along with the tops and windrowed gathered several rows in each row. Generally, it is harvested in March-April in plains and July-september in the hills.

Curing

About 25-50% losses in garlic caused by poor storage conditions. Curing is done to remove excess moisture, to allows bulb to be compact and rest in dormant state. It is done by keeping the bulb in shade for 7-10 days either with tops or after cutting the tops, leaving 2.5cm above the bulb to avoid sun injury. It can be done by either placing the bulbs in the floor or wire sacks.

Grading

After curing the bulbs are graded either manually or by running through the grader. The thick necked, splitted, injured or diseased bulbs with hollow cloves are sorted out. Size grading is done after sorting depending on market demand. The different grades are big (30mm dia), medium (25mm) and small (12mm). After that they are packed in open meshed jute bags of usually 90 and 40-60 kg packs.

Storage

For 6-8 month at the R.H of 70% bulbs are generally stored at any temperature but during this period the bulbs might get damaged by mould formation or sprouting. The ideal cold storage temperature for storing bulbs for 6-8 months is 0-2.2°C with 60-70% R.H. For prolonging the storage life of the bulbs upto 300 days and to check sprouting, pre-harvest application of Maleic Hydrazide 2500-3000 ppm, 3 weeks prior harvest is advisable. Irradiation of bulbs with low dose of 6 krad Cobalt 60 gamma rays' checks sprouting.

Seed Production and Seed Certification Standards

A. Seed Production:

- Isolation distance of 5m should be maintained to avoid mechanical mixtures between two cultivars

- Selection of big sized, healthy bulbs and separate cloves for seed production
- Plant spacing should be same as that of the commercial crop
- All other cultural operations should be similar to that of a commercial crop raising
- Roughing off the diseased and off type plants should be done based on foliage characters
- Harvesting, curing, sorting of bulb should be done in similar way as that of the commercial crop

B. Seed Certification Standard

- **Field inspection:** Minimum two inspections to be made: 1. To the done when the plants are large enough to confirm isolation, off types including bolters and other factors. 2. To be done at final stage when leaves begin to fall and before harvesting of bulbs to check off types, isolation distance and other factors.
- **Field Standard:**

a. **General requirement:**

Isolation: Seed field of the garlic should be isolated from contaminants mentioned below:

Contaminants	Minimum distance	
	Foundation	Certified
Field of other varieties	2 M	3 M
Field of same varieties not confirming to varietal purity requirement for certification	5 M	5 M

B. **Specific requirement**

Factors	Maximum permitted (%)	
	Foundation	Certified
Off types	0.01	0.02

C. **Seed Standard**

- The average weight and diameter of the bulb should not be less than 25g and 2.5 cm respectively.
- The seed material should confirm to the varietal purity and should be clean, healthy and compact. The bulbs not confirming to varietal characteristics should not exceed 0.10 % and 0.20% respectively.
- By weight the damaged, cut, immature and bruised bulb by worms, insects or slugs should not exceed more than 2%.

Future Prospect

Garlic, a popular culinary and medicinal crop, holds promising future prospects for crop improvement. As scientific advancements continue to unfold, researchers and breeders are increasingly focusing on enhancing the various desirable traits of garlic through targeted breeding and genetic modification. These efforts aim to enhance traits such as yield, disease resistance, flavor, nutritional value, and adaptability to different environmental conditions. By harnessing the power of molecular biology, genetic markers associated with these traits can be identified and utilized in breeding programs to accelerate the development of improved garlic varieties. Additionally, advancements in gene editing technologies like CRISPR-Cas9 provide precise tools for modifying garlic's genetic makeup, enabling the introduction of beneficial traits and the elimination of undesirable ones. The future prospects of garlic involve a combination of traditional breeding methods and cutting-edge biotechnology approaches to create garlic cultivars that not only meet the demands of consumers but also withstand the challenges posed by climate change and emerging pathogens. With continued research and development, the future of garlic holds the potential for improved varieties that offer higher yields, enhanced nutritional content, superior taste, and increased resistance to pests and diseases.

References

Bradley, K., Rieger, M., Collins, G. and Armstrong, J. (2001) *Acta Horticulturae*, 555 : 159-160.

Candolle, De, A. (1985) Origin of Cultivated Plants, D Appleton, New York, 1985.

Das, A,K., Som, M.G., Sahu, M.K. and Bose, T.K. (1985b) Indian Agriculturist, 29: 183-189.

Dscalov, C. and Popov, P. (1941) *Fundamentals of Vegetable Growing In Bulgaria*, Sofia, p.431.

Etoh, T., Hong, C.J. and Armstrong, J. (2001) *Acta Horticulturae*, 555 : 209-212.

Janick, J. (1979) *Horticultural Science*, Freeman & Co, Son Francisco. pp. 544.

Jang, Y.S., Oh, Y.B., Choi, I.H., Song, Y.S. and Park, J.H. (2000) *J. Korean Soc. Hort. Sci*., 41 : 157-160.

Kehr, A. E. and Schaeffer, G. W. (1976) *HortScience*, 11 : 422-423.

Kuznetsov, A.V. (1948) Agrbiological Studies of Garlic. Phog Krasnadar.

Mathur R.B.L., Arora, P.N. and Prasad, M. (1975) *Indian Journal of Agronomy*, 11:312.

Maurya, K.R., 1987, Progressive Horticulture, Vol 19, pp. 132-134.

Nasir, A., Fatma, G., Neshat, N. and Ahmad, M, A. (2020). Pharmacological and therapeutic attributes of garlic (Allium sativum Linn.) with special reference to unani medicine- A review. *Journal of Medicinal Plant Studies*, 8(3):6-9.

Om, H. and Awasthi, D.N. (1977). *Progressive Horticulture*, 9: 63-67.

Ohsumi, C., Kojima, A., Hinata, K., Etoh, T. and Hayashi, T. (1993) *Theor. Appl. Genet*., 85 : 969-975.

Pruthi, J.S.(1979) Spices and Condiments (2nd ed), National Book Trust of India, New Delhi, pp. 87-99.

Rao, J.M.V. and Purewal, S.S. (1957) Farm Bulletin No 3, ICAR, New Delhi pp. 18-22.

Regel, E. (1975) *Acta Horticulturae*, 3: 266.

Resende, F.V., Souza, R.J. (1998) *UNIMAR-Ciencias*, 7 : 81-88.

Shinde, N.N. and Sonatakke, M.B. (1986) Bulb Crops (In): Vegetable Crops in India (T K Bose and Som MG ed.), Naya Prokash, Calcutta pp. 544-592.

Singh, G.P. and Singh, R.K. (1984) *South Indian Horticulture*, 32: 127-33.

Tayagi, H. and Aoba, T. (1976) *Yamagata Agriculture and Forestry Society*, 33: 39-50.

Tahir, Z., Saeed, F., Nosheen, F., Ahmes, A. and Anjum F.M. (2022). Comparative study of nutritional properties and antioxidant activity of raw and fermented garlic. *International Journal of Food Properties*. 25(1): 116-127.

Tiwari, J.P., Awasthi, D.M. and Kanuijia, J.P. (1984) *Progressive Horticulture*, 16: 199-210.

Vavilov, N.I. (1926) *Trudy Prikl Bot Genet Sel*, 20:1.

Xiong, Z.Q., Li, S.J., Liu, G.Q. and Huang, B.J. (2000) *J. Nanjing Agri. Univ.*, 23 : 25-28. Nabulsi, I., Al-

Xu, P.W., Srinives, P. and Yang, C.L. (2000) *Thai J. Agri. Sci.*, 33 : 11-20.

10

Advances in Production Technology of Leek

Sajeel Ahamad*[1], Amit Kumar[2], Chhail Bihari[3], Maneesh Kumar[4], Ganesh Kumar Choupdar[5] Menaka M[6] and Vivek Saurabh[7]

[1,5,6,&7]*Ph.D. Research Scholar, Division of Food Science and Postharvest Technology, ICAR- Indian Agricultural Research Institute New Delhi 110012*

[2]*Ph.D. Research Scholar, Department of Fruit Science, Sardar Vallabhbhai Patel University of Agriculture and Technology, Meerut-250110, Uttar Pradesh, India*

[3]*Ph.D. Research Scholar, Department of Horticulture Vegetable Science, Nagaland Central University, SAS Medziphema, Nagaland, India*

[4]*Ph.D. Research Scholar, Department of Horticulture, G.B. Pant University of Agriculture & Technology Pant Nagar, U.S. Nagar, Uttarakhand, India*

[*] *Email: sajeelahamad04@gmail.com*

Introduction

The leek (*Allium ampeloprasum var. porrum* L.) belongs to the Alliaceae family and is a non-bulb-forming crop. It is consumed raw by itself, combined with other foods in salads, and cooked as flavoring in soups and stews. It is a preferred vegetable in a kitchen garden but is not commercially farmed in India. Leek has chromosome 2n=4X=32, which makes it an autotetraploid in nature. It is a vigorous biennial herbaceous cultivar that has never been discovered in its wild form. For their blanched stems and leaves, leeks are grown. It is a biennial plant that is used to flavor soups and eaten as a salad. The leek was first domesticated in the Eastern Mediterranean, where it was later brought to middle Europe by the Greeks, Romans, and Egyptians. Leeks may have originated in Central Asia and/or Europe, according to several theories. Since the majority of the Allium species are found worldwide, their distribution within the northern hemisphere is not uniform.

Based on fresh and dry weight, leek contains more protein and minerals than onion. It has a higher energy content per 100 g of edible part than an onion. Leek is a monocotyledonous plant with small, white bulbs, broad leaves, and an etiolated pseudostem made of the leaf sheath. In contrast to the crop cultivated in the winter, leeks grown in the summer have lengthy pseudostems. While leeks harvested in the winter produce modest stem growth, those harvested in the summer have long pseudostems. Although leeks do not develop bulbs, they are more resistant to the cold than onions (Rana and Pradeep 2017).

Nutritional value

The leek, or *Allium ampeloprasum* L., is one of the most widely grown vegetables in the world and is a member of the Alliaceae/Amaryllidaceae family. Based on their various cytogenetic and morphological traits, the *Allium ampeloprasum* L. is divided into wild leek, cultivated leek, kurrat, pearl onion, and great-headed garlic. Leeks are renowned for their distinct flavour and therapeutic benefits. Leeks are a rich source of phytochemicals and bioactive substances, and they have good nutritional qualities. Dietary fibres, fructans, polyunsaturated fatty acids (PUFA), primarily linoleic acid, different amino acids, primarily glutamine, glutamic acid, aspartic acid, arginine, and alanine, as well as organic acids like ascorbic acid and malic acid are all abundant in leeks. Numerous health advantages are brought about by bioactive substances and phytochemicals such s-alkyl-l-cysteine sulfoxides, polyphenols, saponins, kaempferol glycosides, -carotene, pectic polysaccharides, and tocopherols. These phytochemicals have strong anticarcinogenic and great antioxidant capabilities that contribute to a decrease in colorectal, stomach, and breast cancers. Leeks assist prevent cardiovascular disorders by lowering the risk of hypercholesterolemia, blood pressure, arteriosclerosis, and platelet aggregation. In addition to these, leeks are also effective against a variety of bacteria, fungi, and viruses (Shelke *et al.,* 2020).

Table 1: Chemical composition of leek (per 100 g fresh weight)

Constituents	**Content**
Energy (kcal)	77
Vitamin-A(IU)	30
Moisture (g)	78.9
Thiamine(mg)	0.23
Protein (g)	1.8
Calcium(mg)	50
Fat (g)	0.1
Iron(mg)	2.3
Carbohydrates(g)	17.2
B-carotene (µg)	600
Thiamine (B1) (µg)	120
Nicotinic acid (µg)	500
Pyridoxine (B6) (µg)	250

Constituents	Content
Ascorbic acid (mg)	25
Phosphorous (mg)	30

Source: Van der meer and Hanelt, 1990

Botany

The bent cotyledon with attached testa begins epigeal germination at the apex. Following the sprouting of distichously alternating leaves with tubular sheaths, the first genuine leaf appears through a pore in the cotyledonary base. They combine with young blades to form a white pseudostem that is 15–60 cm long and whose diameter keeps expanding as the plant grows.

Table 2: Major *A. ampeloprasum* crops and main cultivation regions

Crop group	Botanical name	Main cultivation region	English name	Ploidy level
Leek	*A. ampeloprasum* var. *porrum* (L.) J. Gay, *A. porrum* L.	Indonesia, Turkey, Korea, Europe, Kazakhstan	Leek	4x
Kurrat	*A. kurrat* Schweinf. ex Krause	Egypt and adjacent areas	Kurrat, salad leek	4x
Pearl onion	*A. ampeloprasum* var. *sectivum* Lued.	Atlantic and temperate Europe	Pearl onion	4x
Great-headed garlic	*A. ampeloprasum* var. *holmense* (Mill.) Aschers. et Graebn.	California, eastern Mediterranean	Great-headed garlic, elephant garlic	6x
Tarée	*A. ampeloprasum* ssp. *persicum*	Iran	Tarée Irani/ Persian leek	4x

(Fritsch and Friesen, 2002; Dashti *et al.,* 2003, 2005; Mousavi *et al.,* 2006; Guenaoui *et al.,* 2013).

Fig. 1: Botanical description

Table 3: Botanical classification of Leek (Shahrajabian *et al.*, 2021)

Kingdom	*Plantae*
Clade	*Angiosperms*
Clade	*Monocots*
Order	*Asparagales*
Family	*Amaryllidaceae*
Subfamily	*Allioideae*
Genus	*Allium*
Species	*Ampeloprasum*

Taxonomy

The genus *Allium* comprises ~1000 species (Govaerts *et al.,* 2021) including 25 *Ampeloprasum* spp. (4x, 6x and 8x) group/complex (Von Bothmer, 1970; Guern *et al.,* 1991; Mathew, 1996), formed by three or four gene pools. Taxonomically, A. sativum is closer to *A. ampeloprasum* than *A. fistulosum* and *A. schoenoprasum* (Guenaoui *et al.,* 2013).

The domesticated *A. ampeloprasum*: leek, great-headed garlic, pearl onion, prei-anak, Tarée Irani/Persian leek, and mushuu-ninniku (Kik *et al.,* 1997; Hanelt, 2001; Ariga *et al.,* 2002; Hirschegger *et al.,* 2006; Mousavi *et al.,* 2006), were subgrouped as follows:

1. The tetraploids leek (*A. ampeloprasum*); kurrat (*A. kurrat*); Tarée Irani; pearl onion (*A. ampeloprasum* pearl onion group) aka bulbous leek. No evidence for natural hybridization in leek and kurrat was published, while introgression events were identified in pearl onion (Hirschegger *et al.,* 2010).
2. The allopolyploid 6x–8x great-headed garlic (GHG, *A. ampeloprasum* var. *holmense*) (Van der Meer and Hanelt, 1993), aka elephant garlic, giant garlic, English leek, European leek, great round-headed leek, Levant garlic and wild leek (Khazanehdari and Jones, 1997; Fritsch and Friesen, 2002; Brewster, 2008). argued that GHG is a separate taxon from other *A. ampeloprasum* cultivated forms, whose allopolyploid origin includes *A. iranicum* and *A. polyanthum* (Hirschegger *et al.,* 2010).

Soil and climate

The most essential prerequisite for leek cultivation is a loose texture, which may be found in almost all soil types. Peat soil typically produces large yields but poor-quality crops. The best soil types for leek cultivation are sandy clay soils. Heavy soils make harvesting in the autumn and winter challenging. Long white shafts must first form, which requires deep ploughing. The ideal soil has a pH between 6.5-7, is well-aerated, has high drainage and moisture retention capabilities, and should never be flooded (Randy Baker, 1991). In order to generate a longer shaft, deep ploughing is advised. They can be grown well on mineral soils, however harvesting should be delayed and excess soil from the roots may need to be removed by hand more frequently in heavy clay soils. In a chilly to mild climate, leeks thrive. The Oregon coast and Willamette Valley provide excellent locations. They are able to be grown all year long.

Commercial cultivars

Leeks can be divided into four basic categories according to the season in which they mature: (i) summer leeks; (ii) autumn leeks; (iii) autumn and winter leeks; and (iv) winter leeks. The final result is affected by the major differences in growth habits across leek cultivars. According to Randy Baker (1991), they range from long, green, narrow-leaf varieties with long, thin white stems to long, wide-leaf forms with thicker, shorter white stems and blue-green leaves. Leek is a monocotyledonous plant that grows slowly. Leek cultivars differ from one another primarily in traits like vigour, ease of bolting, resilience to cold, leaf spacing, breadth and colour of the leaf blades, and length and diameter of the sheath portion. The most robust varieties are preferable for summer output; resilience to cold is especially important when leeks must be gathered throughout the winter; and slowness to bolt allows a longer harvest period in the spring.

Main season varieties	**Late season - winter**	**Overwinter**
August to October	October to December	Fbruary- March
American flag, jolant, kilima, king richard, primor.	Derrick, electra, goldina, goliath, kilima, tivi, wintereuzen.	Carina, Conqueror (moderate bulbing), Eskimoo & Siberia.

Characteristics of some important varieties

PPL-1

It is a selection from exotic germplasm.

The whole plant is consumed.

Leaves light green, swollen stem and do not form bulb.

Ready in 150-160 days and good replacement for green onion.

Palam Paushtik

An alternative of green onion, suitable for salad, soup and cooking.

Matures in about 140-150 days with average yield of 300-350 q/ha.

Seed sowing

After an approximate 12-week nursery period, leek plants are transplanted. Leeks are always grown from seed and need a considerable amount of time to grow to a size that is suitable for the market. Leeks are commonly transplanted from hotbeds or cold frames into the open as soon as the soil warms up in cool climates. Leeks are typically planted as early in the spring as feasible in these regions. Such early plantings are frequently ready by late summer, and subsequent later plantings may be harvested into the following growing seasons: early fall, early winter, and early spring. Leeks are sufficiently resistant to cold that harvesting can go on all winter long, even in frigid climes, when many other greens are unavailable (Wurr *et al.,* 1999; Jones and Mann, 1963). In plains, August to October is the best period to sow, and seedlings should be planted when they are 15 cm tall. It is sown in hills between March and April. Leeks are more resistant to extreme temperatures than onions. In the growth of the seed stalk, temperature matters more than day length. It does well at higher elevations. Higher altitudes in India are used to cultivate seeds.

Seed rate and seed treatment

Temperature affects seed germination, with 18 to 22°C being ideal. In India, seeds are sown in nursery beds from August to October, and seedlings are prepared for planting when they reach a height of 15 cm. To grow seedlings for one hectare of planting, about 5-7 kg of seeds are required. Seed priming, which involves carefully hydrating seeds so that pre-germination metabolic activities can happen without radicle emergence, can enhance germination performance and field uniformity even more. Unless stored in optimum conditions, the vitality of leek seeds is relatively limited less than two years.

Seeding and transplanting

Leek seeds are often sown directly into fields at a rate of 10-15 seeds per 30 cm row. Seedlings that have emerged are spaced 10 cm apart. Leeks are also transplanted, which is frequently required to have harvests from mid-summer through autumn. Leeks can also be planted directly or transplanted into 15 cm-wide by 15 cm-deep holes. The ditch is filled in as the leek plant expands. Long white stems are produced as a result, which is a desirable quality for leek marketing. The sensitive white section of the leaf stem gets longer the deeper the leeks are trenched or hilled (Decoteau, 2000). For leeks with minimum dimensions of 20 mm and 150 mm, a planting density of roughly 30 plants per square metre is ideal for early production. However, densities of 20–25 plants/m2 are required to yield giant leeks (Brewster, 1994). Another way to transplant is to place each seedling into a hole that has been dug out of the ground and is 15 cm deep. Although this approach does away with the necessity for furrow filling, seedlings should still be placed 40 x 10 cm apart. Due to the lack of registered herbicides and the length of time required to harvest direct-seeded crops (8–12 months or more), direct field sowing is feasible but not advised. Normally, leeks are transplanted. Early in the growing season (February to March) or in outdoor seedbeds (April to May), they should be sown under cover. Leeks for the August–September crop should be sown between mid–February and mid–March. From mid-March to mid-April, seed the crop for early winter harvest. From mid-April until mid-May, seed can be sown for the spring harvest. For each harvest season, the appropriate varieties (above) must be employed. Leek seedlings should be transplanted into 15 cm-deep furrows. During cultivation, furrows are filled in, and as the plants grow, the rows should be hilled up. Leeks should have a white stem that almost reaches the leaf's base. Be careful not to allow soil between the leaves as it will penetrate the shaft.

Nutritional requirement

Nitrogen fertilisation causes bulbs to grow larger in both diameter and length (McCollum, 1976). With a preplanting treatment of 200 kg of n/ha in irrigated and non-irrigated circumstances in dry years, Kaniszewski (1986) recorded the maximum yield of leeks. Split applications of 600 kg of n produced the highest yields during wet years. According to Randy Baker (1991), leeks need between 200 and 250 kg of N (nitrogen) per hectare, preferably in three installments: one third as a side dressing before planting, one third as a top dressing after the leaves have dried, and one third as a pre-plant integrated amount. Leeks don't have very high phosphorus needs; applications of 50–100 kg P2O5 per acre are sufficient. Potash needs are also minimal, and 150–200 kg K2O per hectare as potash sulphate are sufficient in most cases, combined with 80–100 kg n, 60 kg P2O5, and 80 kg K2O/ha of well-rotten farmyard manure. When planting, the soil should be thoroughly mixed with the entire amount of FYM, P2O5, K2O, and half N. One month later, remaining N is side clothed.

Use of growth regulators for proper growth and development

Ethephon @ 50 mg/l as foliar spray should be applied 20-25 days after sowing and NAA @50 mg/l should be applied at 60 and 90 days after planting.

Irrigation

Leeks must grow continuously to be of high quality, and in locations that experience moisture stress, irrigation is frequently required. According to Randy Baker (1991), depending on the weather, a post-planting irrigation is preferred to ensure quick establishment. If there is insufficient rainfall during the hot summer days when significant growth should occur, additional irrigation will be required. To promote robust, uniform growth and sensitive stalks, evenly distribute irrigation. Depending on the species, planting date, and seasonal variations, a total of 12 to 15 inches of water may be needed. Although the frequency of water application is determined by soil type, the total amount of water required is not affected. The lighter soils require less water applied per application but more frequent watering.

A white plastic soil mulch produced larger yields than a black one in leek trials (cv. Santina), but it had no effect when compared to bare land. The period during which overwintering leek crops are marketable is shortened by spring bolting. After then, it takes spring-planted crops around three months to reach marketable size. Transparent crop covers can help accelerate early harvests. Mulches made of non-woven polypropylene textiles or polyethylene film with 500–1 cm diameter perforations per square metre have advanced harvests, according to tests. After a sowing in glasshouses in January, the crop is transplanted and the films are applied at that time, which is typically in late March or early April. These light films are raised by the foliage and appear to "float" on top of the leaf canopy as the crop matures. The perforations are penetrated by rain. Under these mulches, mean temperatures are increased by 1-2°C, causing quicker development. About seven weeks after transplantation, the mulch is removed.

Weed management

The pre-emergence application of ramrod at 7 kg/ha and the post-emergence application of afalon at 1 kg/ha at the 2-3 leaf stage produced the best results in the development of leek seedlings. Dicotyledonous weeds that grow annually could be controlled using just afalon. Both dicotyledonous and monocotyledonous weeds were eliminated by the use of prometryne at a rate of 3 kg/ha after transplantation. Afalon at 2 kg/ha and butisan at 10 l/ha produced satisfactory results, with the former being most efficient against dicotyledons and monocotyledons. Leeks were tolerant of all of these herbicides (Velev and Ivanov, 1973). Inherent traits of onions, such as their slow growth, tiny stature, shallow roots, and lack of dense foliage, make them more susceptible to weed competition than other crops. Identification of plant flora, weed control techniques, and judicious use of effective weed control techniques are all necessary for effective weed management. In India, hand weeding onions is a popular practise, however because of their close spacing and shallow root systems, it is a laborious, expensive, and time-consuming task. Hand weeding becomes challenging during the crop's vital phase due to a lack of labourers, which results in significant output losses. Pre-emergence herbicide application maintains the crop weed-free in the early stages. Later on, the second flush of weeds will have an impact on how bulbs form. Throughout the crop growing stage, hand weeding aids in keeping the weed population below the economic threshold level. Due of the lower workforce availability and greater labour costs, pre-emergence paired with hand weeding may be expensive. Manual or mechanical weed-control techniques used after bulb formation will harm the bulb. Early post-emergence application may be useful to lessen weed competition, bulb damage, and weeding costs. The quality of the leek, however, was reduced (Baumann *et al.,* 2000)

Blanching

In the cultivation of leeks, blanching is crucial. In order to bleach the plants and improve the crop's quality, it is done by covering the plants to a specified height. To achieve this, plants are placed in pits or trenches that have been highly manured, allowing them to grow up to the center of their leaves and eroding the soil. When the plants are young, caution should be taken not to dirt up the soil too soon.

Intercropping

The crucial period for weed control in the intercrop was significantly reduced when leek and celery were intercropped in a row-by-row replacement pattern as opposed to a leek pure stand. In the intercrop, there was a 41% decrease in the relative soil cover of weeds that appeared at the conclusion of the critical time. Senecio vulgaris was planted 20 days after crop establishment in a different experiment, and the biomass of the plant as well as the number of seedlings that emerged as progeny were both reduced when compared to the pure stand of leek. Because of the intercrop's superior resource exploitation, its relative yield total was 10% higher than that of the leek and shallot 375 pure stands.

Disease management

White-tip

Phytophthora porri foister is the root cause of white-tip disease, the most significant leek disease in Europe throughout the winter. Papery-white local lesions with occasionally dark-green water-soaked zones can be seen on infected leaves. There may be significant harvest losses; in some situations, the whole crop may have been lost. According to De Clereq and Bockstaele (2002), a two-step chemical control technique is possible. Maneb is taken as a preventative during the months of September and October, while more systemic medications such benalaxyl or metalaxyl are used as a cure once the first symptoms appear.

White rot (*Sclerotium caplvorum*)

According to Randy Baker (1991), if this soil-borne fungal disease exists in agriculture soils, it can be devastating. The fungus can persist in the soil for a very long time as sclerotia. On ground that has not recently produced a crop from the onion family, leeks should be cultivated. In order to stop the disease from spreading, it is crucial to sanitise field equipment and dispose of cull leeks away from producing areas. The initial symptoms are a yellowing and withering back of the leaves, which start at the tips and move down. Young plants wilt and collapse, and they fall to the ground easily. When this happens, a dense, white mass of mycelium with tiny, black sclerotia imbedded in it is revealed. The growth of white rot is favoured by cool, moist growing seasons.

Leek rust

On European leeks, leek rust (*Puccinia porri* Wint., also known as *P. allii* f. Rudolphi), seriously damages the plant. The uredo stage of leek rust is prevalent year-round because the crop is now grown continuously. The bodies that create endospores, one of the five different types of rust spores, are called uredosori, and they are prevented from forming during the winter by the low temperatures. The leek rust pandemic recurs each spring as soon as the temperature rises. High humidity and little precipitation favour the development of the disease, while exposure to water makes the spores less viable. Temperatures above 24 °C and below 10 °C impede infection while 100% relative humidity (RH) at 10-15 °C results in the highest infection effectiveness. Due to the vulnerability of all leaves to damage and the impracticality of leaf removal, the economic threshold for leek rust is low. According to Schwartz and Mohan (1995), consistent application of preventative fungicides like Maneb or Zineb should provide sufficient defence. Fenpropimorph can be sprayed either by itself or in combination with maneb for effective control. Tebuconazole and epoxiconazole, two triazole-group compounds, are also useful; propiconazole treatments produced excellent control.

Insect pests management

Leek moth

The European native leek moth (onion leaf miner), *Acrolepiopsis assectella* (Lepidoptera: Acrolepidae), a pest of *allium*, was first positively detected in Eastern Ontario in 1993, according to Kristen Callow (2003). The pest can be found throughout Asia, Africa, Europe, and Canada. In some regions of Europe, the leek moth is regarded as a major pest, with infection rates reaching 40% in places where the insect has several generations annually. When there are only one or two generations each year, the pest is sporadic and doesn't do much to harm the economy.

Crop rotation, delayed planting, the removal of old, infested leaves, the destruction of any obvious pupae or larvae, early harvesting (to prevent damage from last generation larvae and population build-up), the positioning of susceptible crops away from infested areas, and the destruction of plant debris after harvest may all be cultural controls that are successful in bringing populations below harmful levels.

Thrips

Thrips (*Thrips tabaci* lindeman) are the main pest of leeks. These 2 mm long insects feed on cell fluids while concealed between the inner leaf blades. As hundreds of highly noticeable grey patches are formed by empty surface cells, the green leaves begin to lose their colour. The cost of quality loss is high economically. When plants are water-stressed in hot, dry weather, thrips damage is at its worst. In these circumstances, leaf expansion is moderate and thrips population growth is quick. The insect develops from an egg to an adult in just 11 days at 30°C (Edelson and Magaro, 1988). Since thrips have a low threshold for harm, extensive study has been done to effectively eradicate this insect. Carbamates, such as furathiocarb and methiocarb, are the superior chemicals. Some phosphorous substances, like pyrethrins and acephate and malathion, may have protective properties.

Harvesting and yield

The crop is ready for harvesting when the blanched basal section of the leaves is at least 1.25 cm in diameter. However, since the plant doesn't produce a bulb, gardeners frequently wait until the plants are 5 cm in diameter before harvesting the crop. The plants may be undercut by the growers to make harvesting easier, just like with onions and garlic. The roots and all but 5 cm of the green leaf blade are removed after the plants are dug up from the soil, leaving largely the white sheath of the overlaid leaves. Although the plants should be harvested in the spring before growth starts again, they can also be kept in the ground over the winter. The potential output of leeks depends on plant population. A stand with more than 160,000 plants per hectare will result from row spacings of 60 cm and plant spacings of 10 cm. 12 bunches of leeks will be sold in about 3600 cartons if we estimate the harvestable crop at 80% of the original stand. Similar to this, 91 cm between rows and 15 cm between plants will result in a stand with more than 70,000 plants and a harvestable crop output of about 1600 cartons holding 12 bunches (Randy Baker, 1991). Leeks continue to grow gently throughout the autumn instead of bulbing or going dormant. Because of this, the time of harvest is quite flexible and depends on the planting date, market conditions, and leek type. Early August is when little leeks can be sold, and kinds that can withstand frost can be gathered all through the autumn and winter. Leeks can now be harvested mechanically, however the majority of leeks are lifted or excavated mechanically before being picked, cleaned, and packed by hand. Harvesters with one or more rows can be utilized successfully. A hectare of leeks produces about 925q green bulbs.

Post-harvest handling

After lifting, the plants are assessed for length and diameter, the outer leaves are removed, the remaining leaves are clipped, they are cleaned or brushed, and then they are packed into boxes. Leeks can be purchased both loose and pre-packaged in plastic bags or trays with covers. Leeks must have the white portion of the pseudostem cut to uniform lengths in order to be pre-packed. Additionally, new items like "baby leeks" are starting to show up in European marketplaces. In Europe, around 90% of the leek production is sold fresh, and the other 10% is processed. Leeks that have been processed can be frozen, freeze-dried, or utilised to make recipes that are already cooked. According to Decoteau (2000), harvested leeks are kept fresh via hydrocooling, icing, or vacuum cooling. Leeks are frequently wrapped in vented polyethylene to prevent desiccation when vacuum cooling is utilised. Leeks can be kept for two to three months at a temperature of around 6 °C and a relative humidity of about 90%.

Packaging

Leeks are frequently cut to a length of 12 inches, bunched according to diameter, and frequently put in polyethylene film bags. They are typically stored in wire-bound crates or 5 kg cartons that hold 10 500 g film bags apiece. Other boxes might be packed with 18–24 bunches, each weighing up to 15 kg in net weight.

Storage

Although leeks are accessible all year long, they are more plentiful from the end of the summer until the beginning of the fall. In the refrigerator, where they will stay fresh for one to two weeks, fresh leeks should be kept unwashed and untrimmed. They will keep moisture if you wrap them loosely in a plastic bag. Leeks that have been cooked are extremely perishable and will only remain fresh for two days even when kept in the refrigerator. After being blanched for two to three minutes, leeks can be frozen, but some of their desirable flavour and texture characteristics will be lost. For around three months, leeks can be stored in the freezer (Anon., 2005a). The quality of the freshly harvested, unwashed leeks was improved when they were kept at 0 °C and 94–95 percent relative humidity in an environment with 2-3% oxygen, 5–10% carbon dioxide, and 96%. 1989's Goffings and Herregods. Leeks should be stored successfully at 0oC for two to three months if handled correctly. After harvest, leeks need to be quickly cooled down to close to 0°C using a vacuum, crushed ice or hydro cooling. They need to be kept at this temperature for the duration of storage with a high relative humidity. Crushed ice and polyethylene film box liners can both be used to stop moisture loss. The elongation and curvature that develop in leeks at 10°C or 21.1°C will be slowed down by proper cooling. Leeks' heat evolution or respiration occurs roughly eight times more quickly at 21.1°C than it does at 10°C.

Future Prospects

The most recent and exciting advancement in leek development occurred about 20 to 30 years ago with the creation of F1 hybrids, however, the adoption of significant biological and genetic breakthroughs lagged far behind. There is definitely room for improvement in seed physiology studies, seed quality, and seed production; CMS is probably not used in commercial seed production; progress in quality, stronger demand for cold units for bolting, tolerance for tropical conditions based on the variety of subtropical lines, and seed production are limited to nonexistent.

References

Anon. (2005c), 'plants for a future: database search results', available world wide web: http://www.ibiblio.org/pfaf

Ariga, T., Kumagai, H., Yoshikawa, M., Kawakami, M., Seki, T. et al. (2002) Garlic-like but odorless plant *Allium ampeloprasum* Mushuu-ninniku. Journal of the Japanese Society of Horticultural Science 71, 362–369.

Baumann, D. T., Kropff, M. J., & Bastiaans, L. (2000). Intercropping leeks to suppress weeds. *Weed Research (Oxford)*, *40*(4), 359-374.

Bothmer, R. V. (1970). Cytological studies in Allium. I. Chromosome numbers and morphology in Allium sect. Allium from Greece. *Botaniska Notiser*, 518-50.

Brewster, J. L. (1994). Onion and other vegetable alliums. Cab. *International Cambridge*, 93-115.

Buttner, R. (2001). *Mansfeld's encyclopedia of agricultural and horticultural crops: except ornamentals* (Vol. 1, pp. 1660-1661). P. Hanelt (Ed.). Berlin: Springer.

Dashti, F., Kashi, A., Vazvaee, A., Mousavi, A., & Ershadi, A. (2005). Study on genetic diversity in genotypes of Taree Irani (*Allium apmeloprasum* ssp. Persicum) using RADP marker. *Journal of Institute of Water, Soil and Plant in Agriculture*, *5*(3), 1-11.

De Clercq, H., & Van Bockstaele, E. (2002). 18 Leek: Advances in Agronomy and Breeding. *Allium crop science: recent advances*, *438*, 431.

DECOTEAU D R (2000), Vegetable Crops, New Jersy, Prentice Hall, Inc

Edelson, J. V. (1988). Development of onion thrips, Thrips tabaci Lindeman, as a function of temperature. *Southwest. Entomologist*, *13*, 171-176.

Fritsch, R. M., & Friesen, N. (2002). Evolution, domestication and taxonomy. In *Allium crop science: recent advances* (pp. 5-30). Wallingford UK: CABI publishing.

Goffings, G., & Herregods, M. (1988, August). Storage of leeks under controlled atmospheres. In *International Symposium on Postharvest Handling of Fruit and Vegetables 258* (pp. 481-484).

Govaerts, R., Kington, S., Friesen, N., Fritsch, R., Snijman, D. A., Marcucci, R., ... & Brullo, S. (2005). World checklist of Amaryllidaceae. *Facilitated by the Royal Botanic Gardens, Kew, 2020*.

Guenaoui, C., Mang, S., Figliuolo, G., & Neffati, M. (2013). Diversity in Allium ampeloprasum: from small and wild to large and cultivated. *Genetic Resources and Crop Evolution*, *60*, 97-114.

Guern, M., Le Corff, J., & Boscher, J. (1991). Comparative caryology of the Allium ampeloprasum complex in France [microevolution]. *Bulletin de la Societe Botanique de France Lettres Botaniques (France)*.

Jones H A and Mann L K (1963), Onions and Their Allies. Botany, Cultivation and Utilization, London, Leonard Hill (Books) Ltd.

Kaniszewski, S. (1986). Effect of irrigation and fertilization on the yield and nutrient status of leek. *Biuletyn Warzywniczy*, *26*(1), 95.

Khazanehdari, K. A., & Jones, G. H. (1997). The causes and consequences of meiotic irregularity in the leek (*Allium ampeloprasum* spp. porrum); implications for fertility, quality and uniformity. *Euphytica*, *93*, 313-319.

Kik, C., Samoylov, A. M., Verbeek, W. H. J., & Van Raamsdonk, L. W. D. (1997). Mitochondrial DNA variation and crossability of leek (*Allium porrum*) and its wild relatives from the Allium ampeloprasum complex. *Theoretical and Applied Genetics*, *94*, 465-471.

Kristen Callow (2003), 'The Leek moth. An introduced pest of Allium present in Ontario and Quebec', Ministry of Agriculture, Food and Rural Affairs, Ontario, Canada.

Mathew, B. (1996) Review of Allium Section Allium. Royal Botanic Gardens, London.

Mccollum G D (1976), Evolution of Crop Plants (NW Simond edn), London and New York, Longman

Mousavi, A., Kashi, A., Davoodi, D., & Shariatpanahi, M. S. (2006). Characterization of an Allium cultivated in Iran: the Persian leek. *Belgian Journal of Botany*, 115-123.

Rana, M. K., and Pradeep Kumar Jatav. "Leek." *Vegetable Crops Science*. CRC Press, 2017. 29-36.

Randy Baker (1991), 'Leek production', Available: Crops Home Page www.omaf.gov.on.ca

Schwartz, H. F., & Mohan, S. K. (1995). Compendium of onion and garlic diseases. St. Paul, Minnesota.

Shahrajabian, M. H., Sun, W., & Cheng, Q. (2021). A Review of Leek (A. ampeloprasum L.), an important vegetable and food ingredient with remarkable pharmaceutical activities. *Pharmacognosy Communications*, *11*(1), 9-12.

Shelke, P. A., Rafiq, S. M., Bhavesh, C., Rafiq, S. I., Swapnil, P., & Mushtaq, R. (2020). Leek (Allium ampeloprasum L.). *Antioxidants in Vegetables and Nuts-Properties and Health Benefits*, 309-331.

Van Der Meer Q P and Hanelt P (1990), 'Leek (allium ampeloprasam)', in brewster J L and Rabinowitch H D, onions and allied crops vol. Iii. Biochemistry, food sciences and minor crops, boca raton, florida, crc press, inc, 179–196

Velev B and Ivanov L (1973), 'Possibility of using herbicides in leeks', Gradinarstvo, 15(4), 21–24.

Warade S D and Shinde K G (1998), 'Other Alliums', in Salunkhe D K and Kadam S S, Handbook of Vegetable Science and Technology, New York, Marcel Dekker, Inc, 415–431.

Wurr, D. C. E., Fellows, J. R., Hambidge, A. J., & Fuller, M. P. (1999). Growth, development and bolting of early leeks in the UK. *The Journal of Horticultural Science and Biotechnology*, *74*(1), 140-146.

11

Advances in Production Technology of Carrot

Jagraj Singh[1*], Satya Prakash[2], Rahul Kumar[3], Amit Kumar[4], Pradip Kumar[5] and Prashant Gautam[6]

[1,2&3] Department of Vegetable Science

[4&6] Department of Fruit Science, College of Horticulture, Sardar Vallabhbhai Patel University of Agriculture & Technology, Meerut-250110 Uttar Pradesh

[5] Ph.D. Research Scholar, Department of Horticulture, Dr. Rajendra Prasad Central Agricultural University Pusa, Samastipur-848125, Bihar

[*] Email- bhullarjagraj1313@gmail.com

Introduction

Figure 1.1: Carrot

Carrot (*Daucus carota* L.) is a member of the Umbelliferae/Apiaceae family. Carrots is an important primitive cool season root vegetable crop cultivated in both tropical and temperate climatic conditions. Carrot is a significant winter root vegetable grown throughout India. Carrots are said to have originated in Afghanistan in western Asia, which is the principal source of variation. Carrots were grown in Asia Minor in the 10th and 11th centuries. It arrived in India from Persia in the 13th or 14th century. Its medicinal cultivation originated between 2000 and 3000 years ago. Carrots are a major root crop in European countries such as the United States, China, Japan, and India. Uttar Pradesh, Punjab, Haryana, Bihar, West Bengal, Assam, Karnataka, Tamil Nadu, and Rajasthan are the main carrot-growing states in India. During 2020-21, India produces 1885 thousand metric tonnes of carrot from 108 thousand hectares of carrot production. During 2020-21, Haryana ranked top in both area (19.83 thousand hectares) and production (337.43 thousand metric tonnes) of carrot in India (**Anonymous, 2020**). Carrot possesses anti-diabetic, cholesterol and cardiovascular disease decreasing, anti-hypertensive, hepatoprotective, Reno-protective, and anti-inflammatory effects. Carrot roots are rich in carotene, iron, thiamine, riboflavin, ascorbic acid, niacin, and sugar.

Roots are also high in carbs, antioxidants, minerals, and fibre. They were used to treat stomach ulcers, abscesses, bladder, liver, and kidney issues, to help with delivery, and even as aphrodisiacs. Carrot seed has an essential oil that is used to flavour liquors and dishes, and it also has antimicrobial capabilities. Carrots that are orange in colour are high in beta carotene, a precursor to vitamin A, as well as thiamine and riboflavin. Asiatic varieties have more anthocyanin pigments and less beta carotene, and they may be less nutritious. Green carrot leaves are abundant in protein, minerals, and vitamins and are used as fodder as well as in the formulation of poultry feed. The enlarged fleshy taproot of the carrot is the edible component. Its roots are used in salads, cooked vegetables such as potatoes and peas, juice, halwa, pickles, preserves, and sweets. Kanji is a beverage prepared from black carrot that is supposed to be a good appetiser. It has a wide range of variations for root colour ranging from white to black; however, orange, red, black, and yellow carrots are the most popular and commercially important globally across all climates ranging from temperate to tropical conditions. Orange roots are mostly used for table and freezer applications, while red roots are utilised for table, juice, halwa, and pickle preparation.

Botany

Carrot is a biennial or annual vegetable including a straight or branched stems that grows from a thick fleshy tap root. Carrot is a cross pollinated crop. Carrots are classified into two types: Asian and European. The pinnatifid leaves, and the eatable section is the fleshy tap root, which is made up of an outer cortex (Phloem) and an inner core (Xylem). Inflorescence is a complex umbel that develops during the second phase. Anthesis in a single umbel takes 7-9 days. Individual blooms have white or yellow petals and are bisexual. In most cultivars, the root grows in length first, then in diameter. The tap root reaches its maximum length 12-16 days after sowing. Several factors determine the shape and size of roots. A few causes that cause cracking, forking, and branching of roots are varietal characteristics, undecomposed organic matter, any barrier to its downward growth, and so on. Temperature has a substantial impact on the colour, shape, and development of roots. Carotene accumulates initially in old phloem cells, then in old xylem vessels, and finally in the centre core.

Taxonomy

Daucus is classified into five sections based on morphological and anatomical characteristics: *Daucus* L., *Anisactis* DC., *Platyspermum* DC., *Chrysodaucus* Thell., and *Meoides* Lange. All cultivated carrots, as well as wild carrots found throughout Europe, are members of the species *Daucus carota* L. The majority of wild *Daucus* varieties can be found throughout Asia as well as the Europe, with a few exceptions in America, Australia, and Africa. All wild varieties found in Asia, Asia Minor, Japan, and the United States of America share the same chromosomal number (2n=2x=18) as their European counterparts.

Soil and Climate

Carrot production requires deep, porous, well-drained, friable textured sandy to loamy soils pH ranging from 6.0-6.5. Deep, well-drained light soil can successfully generate the long, thin, skinny roots necessary for the freshest marketplace. The crop grown in humus-rich moist soil has an excessive number of leaves and produces forked and hairy carrots. On the outside, the roots are similarly rougher and coarser. Carrots are very vulnerable to brackish and saline soils. Heavy, stony, compacted, and poorly drained soil is undesirable because it impedes root development.

Carrot is a predominantly cool season crop. Temperatures ranging from 7.2 to 23.9°C are ideal in order to promote the germination of seeds and 18.3 to 23.9°C for optimal root growth. The ideal temperature for root colour development is 15.6-21.1°C. Temperature influences the development of colour and the growth of the root. Shorter and thicker roots are developed at higher temperatures, while longer, slenderer, and paler roots are produced at temperatures below 10 °C. Prolonged durations and warm temperatures may result in intense flavours and tough roots.

Field Preparation

To achieve the desired yield, the soil for carrot production should be sufficiently prepared. As a result, for seeds to grow, the field must provide a loose, friable, deep, and well-drained environment. Although small seeds direct sown root crops require deep and thorough ploughing, appropriate soil loosening is required for improved root development. Prepare the field ahead of time by repeating 4-5 deep ploughings with a disc harrow or cultivator, then levelling after each ploughing.

Commercial Varieties

Carrot varieties are classified into tropical and temperate types.

Tropical (Asiatic) types

Tropical or Asian varieties should not necessitate cold temperatures for flowering, and they develop seeds in North Indian plains conditions. A brief summary of tropical carrot cultivars is given below-

Pusa Vrishti

Developed in New Delhi by ICAR-IARI. It is a first tropical carrot variety that is heat and humidity tolerant and ideal for Kharif sowing. It has red roots and a self-coloured core. It matures in 85-90 days after planting and produces 250 q/ha roots.

Pusa Rudhira

Developed by ICAR-IARI, New Delhi. It has long red roots, a self-coloured core, and an obtriangular shape. The roots mature around 85-90 days after sowing. It can be sown from mid-September through October. Roots become ready for digging in middle of December and produces 300 q/ha roots.

Pusa Asita

This tropical variety is developed by ICAR-IARI, New Delhi. It is the first tropical black-coloured variety with long black roots and a self-coloured core, and it is a late bolter type that may be sown from September to October. This variety's roots mature approximately 90-100 days after sowing. Fresh carrots can be used to make pickles, salad, juice, and kanji. The yield potential for this cultivar is 250 q/ha.

Pusa Kesar

Developed at IARI, New Delhi by crossing Local Red and Nantes. It is appropriate for early sowing from mid-August to early October. Carotene-rich roots with a self-core that is small and deep red in colour. After sowing, roots take 90 to 110 days to reach maturity, with a yield potential of 300 q/ha.

Pusa Meghali

This variety was bred at IARI, New Delhi by crossing Pusa Kesar x Nantes followed by selection. It contains the greatest amount of Vitamin A due to its short tops, smooth roots, self-coloured core, orange flesh, and stumpy to slightly tapered roots. It can be sown as a late crop in October or November or as an early crop in August or September. After sowing, roots reach maturity in 100–120 days, and the estimated yield is 250–280 q/ha.

Pusa Vasuda

The Delhi State Seed Sub-Committee released the first public-sector tropical carrot hybrid generated through the CMS method in 2013. The roots are self-coloured, sleek, attractive, sturdy, red, and juicy, with a high concentration of total carotenoids, lycopene, TSS, and minerals. Suitable for salad, juice extraction, frying, and carotenoid extraction in industry. After sowing, roots reach maturity in 80-90 days, and the estimated root yield is 400 q/ha.

Kashi Krishna (VRCAR- 126)

This variety is developed by IIVR, Varanasi. Attractive black roots with a self-coloured core. Bolting resistant, with less secondary roots/scars, and ideal for salad, juice, halwa, kanji, and nutraceutical applications. Roots are the rich source of anthocyanin (285 mg/100g FW), phenolics and antioxidants. Average root yield is 200-220 q/ha.

Arka Suraj

This variety was developed by IIHR, Bangalore through pedigree selection between Nantes and IIHR 253. Flowers and seeds are set under tropical conditions in this cultivar. Deep orange roots with such a self-colour core and a smooth, conical root surface. The root length is 15-18 cm, and the root diameter is 3-4 cm. TSS of 8-10^0 brix. Tolerant to powdery mildew and nematodes.

Pusa Kulfi

It is the first light mustard-coloured carrot variety developed by ICAR-IARI, New Delhi. It has a marketable maturity period of 90-100 days as well as an average root production of 27 t/ha. Roots are abundant in carotenoids, lutein, thiamine, and niacin.

Selection No. 233

It is a cross between Nantes and No. 29. It was produced by the Punjab Agricultural University, Ludhiana. The roots typically 15 cm long, thin, semi-cylindrical orange, and have a light orange core. It is a late bolting variety that is free of splitting and takes 90-100 days to develop roots.

Selection No. 29

Punjab Agricultural University, Ludhiana, made a selection from locally available materials. It grows quickly, has long, tapering, thin roots that are light red in colour, and has a wide range of adaptability.

Temperate types

The temperate types form roots both under tropical and temperate conditions but set seed only under temperate conditions as they need low temperature of 5 -8^0 C for 40-60 days before throwing out the seed stalks. A brief description of tropical types of carrot varieties is given below-

Pusa Yamdagni

It is a temperate carrot cultivar produced by IARI Regional Station, Katrain. It is a cross between EC- 9981 x Nantes. Orange roots with a self-coloured core, 15-16 cm long, gradually medium tops, tapered or semi stumpy, and high in beta-carotene. It can be sown from mid-October to early December and matures in 90-100 days. It is recommended for cultivation in throughout India. The average root yield per hectare is 90-100 q/ha.

Pusa Nayanjyoti

Developed at the IARI Regional Station, Katrain. It is first orange coloured and self-coloured core temperate carrot F1 hybrid from public sector in India using CMS. It contains a significant amount of beta carotene (7.55 mg/100 g). Average yield of this variety is 350-450 q/ha.

Nantes

The roots are approximately half a foot long, narrow, cylindrical in shape, with a stump and a short thin tail. It has a tasty flavour, an extremely fine texture, and a self-coloured core with orange scarlet flesh. It develops in 90-100 days and is suitable for sowing from mid-October to early December. It gives 120 q/ha of roots.

Early Nantes

The roots have an almost cylindrical shape, a tapering end. The roots are 12-15 cm long and orange with a self-coloured centre. It's ideal for canning. It is suitable for sowing from mid-October until December. The roots become mature in 90-100 days after sowing with 100-120 q/ha root yield.

Nantes Half Long

It is evolved at IARI Regional Station, Katrain. The roots are cylindrical in shape, small, skinny, rough, sweet, and stumpy with abrupt tail, small core and orange in colour along with a self-coloured core. From mid-October to early December, it can be grown. It matures in 90-100 days with a yield of 150-200 q/ha.

Chantenay

The roots of this variety are smooth, half-length, with thick shoulders and gradually tapering to a stump end, with a deep reddish-orange colour and self-colour core. It can be sown from mid-October to early December and is also a suitable cultivar for canning and storage. After sowing, roots take 100 to 120 days to reach maturity. Average root yield is 125-150 q/ha.

Imperator

Imperator is a cross between Nantes x Chantenay. The roots are 15-17.5 cm long, 2.5-4.5 cm in circumference, deep orange, tapering slightly from crown to tip, with a very small core as well as a self-coloured core. It can also be sown from mid-October to early December and matures in 120-130 days. Average root yield is 120-150 q/ha.

Zeno

Introduced from Germany in the Nilgiris hills of Tamil Nadu. Roots of this variety are deep orange with a self-coloured core and mature in 110-120 days after sowing.

Ooty-1

It was developed at the Horticultural Research Station in Ooty, Tamil Nadu Agriculture University. Roots of this variety are deep-orange with self-coloured core. This variety's roots mature approximately 100-110 days after sowing.

Seed Rate and Seed Treatment

About 6-8 kg of seeds of Asiatic types and 5-6 kg of the European types are sufficient for sowing in a one-hectare area. The seeds generally take 8-10 days for germination under ideal climatic circumstances. It is preferable to provide light irrigation immediately after sowing or soaking seeds in water for 12-24 hours prior to sowing can enhance seed germination.

Sowing and Transplanting

Flat beds, raised beds, or ridges can all be used to sow seeds, however ridges or raised beds are preferable for best drainage and to promote root length and smoothness. Tropical (Asian) varieties should be sown on the northern and eastern Indian plains from August to early October, and temperate (European) varieties from October to December. Sowing starts in the hills regions from March to till July. Carrot seeds are sown at a spacing of 25-30 x 8-10 cm. On the ridges, the seed should be put at a depth of 1.0 cm. After germination, thinning the plants to leave 8–10 cm space between each row's plants.

Nutrition Management

Add 200-250 quintal well rotten FYM or compost per hectare during the preparation of field. Fertilizers should be applied based on the soil's fertility level. In normal soils, 120 kg Nitrogen, 60 Kg of Phosphorus and 60 Kg of Potassium per hectare is sufficient for optimum crop production. At the time of sowing, a half dose of nitrogen and a full dose of phosphorus and potassium are supplied at the bottom of the ridges and covered with the soil. Rest half dose of Nitrogen is given to the crop 30 days after sowing. Green manures, vermicompost and biofertilizers like *Azospirillum* and *Phosphate Solubilizing Bacteria* are helpful in increasing the yield and improving the root quality.

Irrigation Management

For enhanced seed germination, growth, and root development, the field must have the ideal amount of moisture. In order to produce carrots, water must be used in the right quantities and at the right times. As a result, soil should never be allowed to completely dry out. Soil moisture is crucial for germination as well as root development. After sowing, the crop should be moderately irrigated, if pre-sowing irrigation is not given to the field. In accordance with the amount of rain or other meteorological circumstances, crops should be irrigated at an interval of 8 to 10 days. In the carrot field, both stress and water stagnation cause unattractive flavour, poor colour development, and, eventually, low-quality roots. Carrots grown under low moisture have an extremely intense and pungent flavour. The water requirement of carrot crop is 300 mm per hectare.

Weed Management

Because carrot is a slow growing root crop, intercultural operations should be executed at an early stage on a regular basis to eliminate weeds. Weeds should not be able to compete with crop during any stage of crop growth. Weeds are manageable. manually, mechanically, chemically, or by a combination of these approaches. In general, two hand weeding at 20-35 days after sowing are adequate to reduce weed growth. Spray pre-emergence weedicide pendimethalin at a rate of 3.25 litres per hectare immediately after sowing whenever soil moisture is sufficient. A pre-emergence spray of Propazine at 1.12 kilogramme per hectare is necessitated for effective weed control. During weeding and hoeing, the crown of the roots is usually exposed to light, which causes greening and decreases the quality of roots. During the root formation stage, earthing up is also required for better root growth and development.

Physiological Disorders and their Managements

Cavity spot

This disorder is distinguished by the cavity development in the cortex, this causes the subtending epidermis to collapse, resulting in a pitted lesion. It is caused by a deficiency of calcium, an oversupply of potassium, and a delay in harvesting.

Control- Add calcium-containing fertilisers to the soil and harvest the crop at right maturity stage.

Carrot Splitting

Splitting of carrot roots is a key physiological disorder of carrots where roots break and seems to be influenced by genetic aspects, but a variety of other aspects such as heavy application of nitrogenous fertilizers in the early stage, low chloride content in the soil, heavy irrigation after a long dry spell, and wide difference in temperature ranges during the later phase of root development may be involved. Splitting may also be an issue with large-rooted cultivars and crops sown in wider intervals.

Control- Use the recommended amount of nitrogenous fertilizer. Sow the seed with adequate spacing. Manage soil moisture in the field and harvest the crop at the proper stage of maturity.

Forking

Roots get forked when the hard soil does not allow straight growth in poorly prepared soil. It is generally caused by mechanical injury to growing roots due to intercultural operations, stones, compact soil and or moisture stress.

Control- Deep ploughing of the soil before sowing may alleviate some of these issues while also assuring a neatly prepared seedbed so that the primary roots may grow freely.

Greening

Greening is a genetic characteristic found only in few cultivars. Greening occurs when the crown is exposed to sunshine. This is more of an issue in light soils that are susceptible to wind or water erosion, as well as when carrots are grown in raised beds.

Control- Earthing-up will completely control the situation.

Major Insect- Pest and their Management

Carrot Rust Fly (*Psila rosae*)

The slender, yellowish-white maggot feeds into the roots, leading them to distort and susceptible to decline, making them unfit for consumption and marketing. Wilting and redness of foliage are the main symptoms above the ground.

Control Measures- Spraying the crop with Diafenthiuron 50% WP at 1g/litre of water or Thiamethoxam 25% WP at 0.5 g/litre of water. Sowing should be done at proper spacing to avoid thinning.

Aphid *(Aphis fabae, Myzus persicae*)

The sap is sucked from the tender sections of the plant by insects. When there is a strong infestation, the leaves get curled, yellowed then eventually die. Their rapid multiplication is aided by the overcast and humid circumstances.

Control Measures- Spray the crop with Imidacloprid 17.80 SL @ 0.5-1 ml/litre of water.

Cutworm (*Agrotis ipsilon, A. segetum*)

Carrots are occasionally infested by cutworms. The majority of cutworms hide underneath the soil surface during the day and emerges to feed at night. Chemical treatments work much better when applied to wet soil in the early evening. Cutworms are aggressive feeders of young foliage and stems, cutting numerous young seedlings in an evening.

Control Measures

Caterpillars are exposed to birds and other predators due to deep ploughing and churning of soil, as well as flooding of fields. Caterpillars spotted just underneath the injured plant were hand-picked and destroyed.

Major Diseases and their Management-

Alternaria blight *(Alternaria dauci)*

The symptoms of this fungal disease begin as a dark brown lesion on the leaf with a yellow edge and spread to the entire foliage drying up and withering under favourable conditions. It infects the stem, pods, and seeds as well.

Control Measure- Before the sowing, the seeds should be treated with hot water at 50^0 C for 30 minutes. Foliar application of Azoxystrobin 20% W/V + Difenoconazole 12.5% W/V @ 0.5-1.0 ml/liter of water should be applied to crop.

Water soft rot (*Sclerotinia sclerotiorum*)

This fungal disease causes carrot deterioration in storage. Affected roots soften and become wet, and white mycelium with black sclerotia forms quickly. It can also infect the roots of plants in the field.

Control Measure- Store the roots at $0\text{-}2^0$ C temperature and select the affected roots and destroy them. Spray the crop with Tebuconazole 25% WG @ 1.5-2.0 g/liter or Carbendazim 12% + Mancozeb 25% @ 1.0 g/liter of water at regular intervals.

Bacterial Soft rot (*Erwinia carotovora*)

This is a destructive bacterial disease of carrots, that occurs during storage and transportation. The infected roots grow soft and watery, and as rooting progresses, they discharge a watery liquid that produces a foul odour.

Control Measure- Bruising should be avoided during harvesting, grading, transportation, storage, and handling. Clean and dry the roots before storing them at 0 ^{0}C temperature and 90% relative humidity.

Yield

The yield varies depending on the type and the growth circumstances. In general, Asiatic/tropical carrots produce higher yields (250-300 q/ha) than European/temperate types (100-200 q/ha).

Harvesting

To assure quality, carrots for the fresh market are harvested before the plants achieve optimal maturity, whilst carrots for processing are allowed to grow for a longer length of time to increase yield. Small, consistent, soft, and mild-flavoured roots should be gathered for the fresh market. When Asiatic carrots reach a marketable maturity level, they are 2.5- 4 cm in diameter at the top.

Post-harvest management

After harvesting, the roots are shaded before being washed, cleaned, graded and in sets of 6 to 12 roots. Under normal conditions, even the freshest carrots can be kept for more than 3 to 4 days. In cold storage, long-term storage is possible without a discernible change in quality. Carrots stored in crates wrapped with perforated plastic film at 0^{0} C and 93-96% relative humidity for 7-8 months show no change in nutritional or organoleptic qualities. Carrots should not be exposed to ethylene gas or ethylene-generating ripening produce during storage or transportation. Ethylene stimulates the development of chemicals that cause carrot bitterness. Browning of roots during storage is caused by moisture loss subsequent harvest. This is due to harvesting of carrot in hot weather, a delay in transferring carrots to storage, and a failure to immediately cool. Carrots must be cooled at 0^{0} C as early after they are harvested to avoid spoilage browning and enhance postharvest shelf life.

Future Prospects

Carrot is a vital vegetable that is grown and consumed all over the world. Its represent one of the most nutritionally rich root vegetables grown all over the world. Carrot cultivation is profitable in the majority of regions. Asia alone accounted for about 60 percent of global carrot production. However, successful carrot production depends on the use of high-quality planting materials and using effective farm management strategies. The adoption of modern production technology, such as improved varieties/hybrids, IPM, INM, efficient farmer training, improved marketing alternatives, and government development policies, resulted in a significant rise in carrot production and productivity in the country. Crop improvement aims to develop cultivars with high root production and carotene level that are suitable for different temperature circumstances, as well as thick fleshy roots with a thin self-coloured core, cylindrical consistently tapering, non-branching, and non-cracking roots. In addition, efforts have been made to critically analyse an overview of research successes in the sectors of breeding programs, production, and performance - based approach, as well as future strategies to encourage carrot research in India. Within distinct populations, breeding techniques include mass selection and pedigree selection. The alternative strategy is hybridization and hybrid development. This is often based on two systems of cytoplasmic male sterility with distinct genetic bases and origins. Carrot population enhancement often begins with inter crosses among open-pollinated variety plants, followed by phenotypic mass selection for root shape, softness, thickness, and ideal root colour in a specific local source of production. Disease resilience, root quality, cavity spot, nematodes, aroma, carotenoid content, and processing quality all need to be addressed. Plants with significant root splitting, early bolting, or weak plant vigour are always eliminated.

References

Anonymous (2020). Area and Production of Horticulture Crops 2020-21, National Horticulture Board, Gurugram.

Fageria, M. S., Chaudhary, B. R. and Dhaka, R. S. (2009). Production Technology of Vegetable Crops. Kalyani Publishers, New Delhi. pp, 117-122.

Gopalakrishnan, T. R. (2007). Vegetable Crops, New India Publishing Agency, Pitam Pura, New Delhi, 4, pp.249-254.

Hazara, P., Chattopadhyay, A., Karmakar, K. Dutta, S. (2011). Modern Technology in Vegetable Production. New India Publishing Agency, Pitam Pura, New Delhi-88, pp. 266-274.

http://ztmbpd.iari.res.in/technologies/varietieshybrids/vegetables/carrot/

https://iivr.icar.gov.in/kashi-krishna-vrcar-126

https://iihr.res.in/carrot-arka-suraj-0

https://www.gov.mb.ca/agriculture/crops/crop-management/pubs/bmz00s05p.pdf

https://seednet.gov.in/cms/Home/Noticeandother/list%20of%20new%20varieity%20of%20 vegetaple%20crops.pdf

https://www.asiafarming.com/carrot-farming-information-guide

https://www.iari.res.in/files/IARI-News/IARI_News_Oct-Dec_2015-15022016.pdf

Laxmi Chhetri, L., Rizwan, M. D., Munkombwe, S., Dorji, N. and Mutum, E. (2022). Utilization and characteristics of black carrot (*Daucus carrot* L.): Potential health benefits and effect of processing. *The Pharma Innovation Journal,* 11(6): 30-41.

Sharma, A., Katoch, V. and Bhushan, B., Practical Manual Olericulture-I, The Secretary, CBSE, Central Board of Secondary Education, Shiksha Kendra, 2, Community Centre, Preet Vihar, Delhi-110092, pp.9-10.

Singh, B. (2015). Vegetable Science, Rama Publishing House, Meerut, 250002, pp. 220-224.

Singh, D. (2019). Vegetable Science, New Vishal Publications, E-153, West Patel Nagar, New Delhi, 110008, 1, pp. 114-119.

Singh, N., Roy, S., Karmakar, P., Chaurasia, S.N.S., Gupta, S. and Singh, B. (2015). Improved Production Technologies in Vegetable Crops. *IIVR Training Manual No. 59*. ICAR-Indian Institute of Vegetable Research, Varanasi, India, pp.128-133.

Thamburaj, S. and Singh, N. (2018). Textbook of Vegetables, Tuber crops and spices. ICAR. New Dehli, pp. 151-156.

Thind, S. K. and Mahal, J. S. (2021). Package of Practices for Cultivation of Vegetables. Additional Director of Communication for Punjab Agricultural University, Ludhiana. pp. 83-85.

Varshney, K. and Mishra, K. (2022). An Analysis of Health Benefits of Carrot. *International Journal of Innovative Research in Engineering & Management*, 9(1): 211-214.

12

Advances in Production Technology of Radish

A. Pranitha Reddy[1], *A. Mamatha*[2], *L. Yashwanth Reddy*[1], *K. Nirosha*[2] *and M. Shiva Prasad*[1]

[1]*M.Sc. Vegetable Science, College of Horticulture, Sri Konda Laxman Telangana State Horticultural University, Rajendranagar, Hyderabad-500030, Telangana*

[2]*Asst. Professor, Department of Vegetable science, College of Horticulture Sri Konda Laxman Telangana State Horticultural University, Rajendranagar Hyderabad -500030, Telangana*

Email: pranireddy40@gmail.com

Introduction

Figure 1.1: Radish

The botanical name of carrot is *Raphanus sativus* which belongs to the family Brassicaceae having haploid chromosome number of x = 9 and is believed to be originated from Europe and Asia. Radish is one of the most ancient and popular leaf cum root vegetables in both tropical and temperate regions. Radish probably originated in Europe and Asia. Radishes were extensively cultivated in Egypt and ancient records show that radishes were eaten before the pyramids were built. It had spread to China during 500 BC and to Japan during AD 700. The recent study has indicated that east of Mediterranean region, most probably China is the centre of origin of radish. Wild state of radish is not known. Previously it was believed to haveevolved from a common weed *Raphanus raphanistrum* Linn in Europe but now it is believed to have been evolved from more a variety of sources. As it is a short duration annual or biennial crop, it can be widely used in sequence cropping, companion cropping and intercropping in widely spaced crops. Roots are pink in colour due to presence of anthocyanin pigment. The characteristic pungent flavour of roots is due to the presence of volatile alkaloid isothiocyanate.

Nutritional composition

Radish is a rich source of Ascorbic acid containing 15 to 40 mg/100 g and has a wide variety of minerals. Pink skinned radish has higher ascorbic acid than white skinned radish. The pungent flavour of radish is due to the presence of volatile isothiocyanates. Major sugar present in radish is Glucose.

Nutritional composition of radish roots, tops and fruits (values are in 100 g edible portion)

Constituents	Radish roots		Radish tops	Radish fruits
	White varieties	Pink varieties		
Moisture (g)	94.4	90.8	90.3	90.5
Protein (g)	0.7	0.6	2.7	2.3
Fat (g)	0.1	0.3	0.6	0.3
Fiber (g)	0.8	0.6	0.9	1.4
Carbohydrate (g)	3.4	6.8	3.4	4.7
Minerals (g)	0.6	0.9	2.1	0.8
Calcium (mg)	50	50	310	80
Phosphorus (mg)	22	20	60	100
Iron (mg)	0.4	0.5	16.1	2.8
Vitamin A (IU)	50	50	18.66	50
Thiamine (mg)	0.06	0.06	0.03	0.07
Riboflavin (mg)	0.02	0.02	0.16	0.05
Nicotinic acid (mg)	0.5	0.4	0.3	0.2
Vitamin C (mg)	15	17	103	69
Energy (kcal)	17	32		

Source: Aykroyd (1963); Gopalan *et al.* (1982); Bose and Som (1986); Hazra and Som (1999)

Uses

Radish commonly known as muli is known for its young tender roots which are eaten raw as salad or cooked as a vegetable. Radish is a common pungent ingredient used in various abdominal disorders and considered as an appetizer. Leaves are also cooked as a vegetable. Radish is useful for liver and gallbladder troubles. They are also used in neuralgic headaches, sleeplessness and chronic diarrhoea in homeopathy. Roots, leaves, flowers and pods of radish are very effective against gram positive bacteria. (Kirtikar and Basu, 1935). All parts of the plant are used in medicine. The seeds are expectorant, digestive, diuretic, laxative and carminative and the root extracts has been reported to have anti-urolithiatic properties. (Alqasoumi *et al.,* 2008).Pocasap *et al.* (2013) reported that the extract of thai rat tail radish showed cancer preventive effect.

Botany

Botany is the scientific study of plants, and the botany of radishes is concerned with the growth, structure, function, and classification of the radish plant. Radishes belong to the family Brassicaceae and are classified as *Raphanus sativus,* Radishes are native to the Mediterranean region but are now grown in many parts of the world. Radish is an annual or biennial depending on its ecotype. Edible portion of radish develops from primary root and hypocotyl. The rosette leaves of radish are pinnatified, lyrate and are in various sizes *viz.,* 10 to 15 cm in cultivars with small roots and 45 cm in cultivars with large roots. Its inflorescence is a typical terminal raceme. Flowers are small, usually white in colour. Sepals are erect and 4 in number. Fruit type of radish is siliqua. The ripe and mature seeds are yellow and as they age, they turn reddish brown. Radish is highly cross-pollinated because of the presence of sporophytic self-incompatibility. It is mainly pollinated by Honey Bees. Bud pollination is used for selfing in radish. Flower buds are pollinated two days before they open by collecting the pollen from the flowers of same plant which are previously bagged.

The radish plant has an edible root, which is usually white with a pink or red exterior, and edible leaves that are often used as a cooking green or in salads. The root of the radish is the main part that is harvested, and it contains a high level of water and a crunchy texture. The radish plant is a fast-growing crop that is relatively easy to grow in well-drained soil and full sun. The seeds are usually sown directly in the ground and germinate quickly, producing a mature plant in as little as four weeks. The leaves of the radish plant are generally green, and the flowers are small and white or yellow, depending on the variety. In conclusion, the botany of radishes encompasses the growth, structure, function, and classification of this important plant species. Radishes are a fast-growing, easy-to-grow crop that is both nutritious and versatile, and have been cultivated for thousands of years for their culinary and medicinal uses.

Taxonomy

According to Banga (1976), there are four types of radish cultivars in various regions of the world are as follows

i) small, cool season radish

ii) large radish with wide range of temperature adaptation

iii) Rat-tail or mougri radish forming no fleshy root but forming long slender (20-60 cm) pods.

iv) Fodder radish also producing no fleshy root. All four types of radish belong to the species *Raphanus sativus* L. with $2n=18$, and botanically they are known as varieties *radicula, niger, mougri (*or*caudatus*) and *oleifera*, respectively. All the four types inter cross freely with each other and also with related wild species.

Cultivated radishes are put under 3 different species based on the plant morphology and ecological requirements.

i) ***Raphanus indicus***: These are Indian types which are heat-tolerant, pungent, and mostly used for pickle- making.

ii) ***Raphanus sativus***: These are European types which are frost-tolerant, rich in carotene, long-day plants, and strictly annuals.

iii) ***Raphanus sativus***: These are Japanese types which are intermediate to Indian and European types, growing in the coastal regions of Japan and neighbouring regions, and turnip-shaped giant radishes cultivated in Sakurajima Island of Japan.

Turnips are classified into the following categories

Rank	**Scientific Taxonomical Details**
Kingdom	Plantae
Subkingdom	Tracheobionita
Superdivision	Spermatophyta
Division	Magnoliophyte
Class	Magnoliopsida
Group	Dilleniidae
Order	Brassicales
Family	Brassicaceae
Genus	*Raphanus*
Species	*Raphanus sativus* L.

Soil and Climate

Radishes can be grown in all types of soils, but the best results are obtained in deep, well drained, light, friable loam soil that contains ample humus. Since it is a short duration crop, it can be grown on soils that are not considered satisfactory for other root crops. For early crop, sandy or sandy loam soils are preferred; however, for summer crop a cool and moist soil gives best result. Usually, heavy soils produce rough, ill-shaped roots with a number of small fibrous laterals, and such soils should be avoided.

Field preparation

The soil should be deeply ploughed and harrowed to make it friable and free from clods. Organic manures are very useful for making the soil fit for development of roots. 20 to 25 tonnes of well rotten farm yard manure should be applied during land preparation about 20 to 25 days before sowing. Amount of farm yard manure may vary depending upon the availability of NPK in the soil.

Commercial Varieties

There are many commercial varieties of radish available, which are developed to meet specific market demands and growing conditions. Some popular commercial varieties of radish include:

Asiatic type varieties

Their crisp, juicy texture and are often used in salads, sandwiches, and as a garnish. Some Asiatic type varieties of radish are characterized by their elongated or cylindrical shape and a crisp, juicy texture. They typically have a white, red, or black skin and a white, red or purple flesh. They are generally larger in size and have a more intense, spicy flavour compared to European type radishes. Asiatic type varieties of radish are typically faster to bolt (forming seed stalks) than European types, which means they are better suited to the summer planting. They are also known for their large size and intense, spicy flavour, making them ideal for use in pickling and stir-fries. These varieties are also popular in Asian cuisine, where they are often sliced thin and eaten raw as a garnish or condiment. In general, Asiatic type varieties of radish are known for their large size and intense, spicy flavour, making them ideal for use in pickling and stir-fries. They also come in a variety of shapes and colours, and are popular in Asian cuisine. These radishes are also known for of the most popular Asiatic type varieties include:

1. **Pusa Chetki**: It is more tolerant to high temperature and can flourish during the hot months. In southern and western parts of India, it can be grown round the year. The roots are pure white, smooth, very soft in texture, mildly pungent in taste, medium long, and thick in size, and almost stumpy in shape. The average length of roots ranges from 12 cm in summer to 20 cm in early monsoon season. Growing period is from April to early September and its best growing period is from July to early September. This variety becomes ready for harvesting in 40-45 days after sowing and yields 20 to 25t/ha.

2.Pusa Desi: It is a derivative from mass selection. Developed for growing during *kharif* season in north Indian plains and can be grown from middle of August to middle of October. Roots are pure white, mildly pungent, 30 to 35 cm long with green stem end. It becomes ready for harvest 50 to 55 days after sowing.	
3. Pusa Reshmi: It is developed at Indian Agricultural Research Institute, New Delhi. It is suitable for sowing in cooler months from mid-September to early October. Roots are white, slightly pungent, 30 to 35 cm long with green shoulder. Crop duration is 55 to 60 days.	
4. Japanese white: It is developed through introduction at IARI Regional Station, Katrain. Suitable for sowing between October to December in plains and September in hills. Roots are white, 20 to 30 cm long, cylindrical, slightly pungent with green shoulder. It yields about 15 to 30 t/ha in 60 to 65 days after sowing.	
5. Punjab Agethi: It is developed at PAU, Ludhiana. Roots are medium sized, red skinned at the top and white skinned at the lower half, less pungent, smooth surfaced with few hair. The best suitable variety for sowing from April to august. The average yield is 37 t/ha.	
6. Punjab safed: It is developed at PAU, Ludhiana from a cross between 5-white and Japanese White. The roots are white with green shoulder, 30 to 45 cm long, mildly pungent, smooth with no hair, tapering into a semi blunt tip and free from forking. Suitable for September to October sowing. It is quick growing and takes 50 to 60 days from sowing to harvesting.	
7. Punjab Pasand: This variety has been developed by repeated crossing of selected plants from a base population of cv.Pusa Chetki. Roots are long, pure white, semi-stumped, free from hair and suitable for sowing in October and November. It is a main season and quick growing variety attaining marketable maturity within 40 days after sowing with a root yield of 53t/ha.	

8. Kalyanpur White: It can be grown all-round the year except during summer months. The roots are pure white, mildly pungent, 25 to 30 cm long, uniform in size with blunt end and white top. Matures in 45 to 50 days after sowing.	
9. Kalyanpur No-1: Roots are white with green shoulders, 22 to 23 cm long and tapering.	
10. Chinese pink: This variety has 30 to 40 cm long shining red and smooth surfaced roots with semi blunt end. The root flesh is white, crispy, mildly pungent and solid. It takes 50 to 55 days for root formation. This variety grows well in plains under mild climate.	
11. Arka Nishant: It is an improvement over a collection IIHR-72 from Singapore developed at IIHR, Bangalore after 10 cycles of mass selection. It has long, marble white, mildly pungent and crisp textured roots which do not develop pithiness. It is resistant to premature bolting, root branching and forking. It sets seeds freely in plains. It is a multiple disease resistant variety of radish.	
12. CO-1: This was developed by TNAU, Coimbatore. Roots are milky white, cylindrical and 23 cm long. It is a short duration variety of 45 days with root yield of 22 t/ha.	
13. Kashi Aardra: Developed from IIVR, Varanasi in 2019. Suitable for sowing during mid-September to mid-December. Roots are attractive white in colour, long and icicle in shape. Sowing time is Rabi. Roots are ready to harvest in 40 to 45 days after sowing. It yields about 24 to 40 t/ha.	
14. UHFR-12-1: It is developed in 2019. Roots are round to slightly tapering, 10 to 12 cm in length and white in colour, weighing about 180 to 200 g at edible maturity. Roots are ready to harvest in 55 to 60 days after sowing. It yields about 35t/ha.	
15.Pusa Gulabi: Entire pink fleshed variety with average yield of 35 t/ha. High total carotenoids, anthocyanins (3.93 mg/100g) and optimal ascorbic acid.	

16. Pusa Jamuni: First purple fleshed variety with average yield of 30 t/ha. Rich in anthocyanins (8.51 mg/100g) and ascorbic acid. Useful in the purification of the blood and eliminate toxins, waste and useful in treating jaundice.	
17. **Kashi Lohit:** It is a variety of radish have roots with an eye-catching red hue and an icicle-like form. They are ideal for use in salad dressings and contain high levels of antioxidants, which are up to 100% higher than those found in white radish. Additionally, these radishes are a good source of anthocyanin and phenolic content and have a low risk of becoming tough or fibrous. The root yield for these radishes is estimated to be between 40-45.0 tons per hectare.	
18. Kashi Mooli-40: The Kashi Mooli-40 variety of radish is well-suited for summer cultivation and can withstand high temperatures (35-42 °C). It is less prone to bolting and has a low risk of becoming tough or fibrous. The roots have an icicle shape and a beautiful white color. Under normal conditions, this radish has a root yield of 30-35 tons per hectare, but this may decrease to 20-23 tons per hectare during the summer season. The recommended seed rate is 5-6 kilograms per hectare.	

European type varieties

European varieties of radish are characterized by their round or oblong shape, smooth skin and a mild, slightly spicy flavour. They typically have a white or light-colored flesh and are relatively small in size, with a diameter of 2-3 inches. European type radishes are typically slower to bolt (forming seed stalks) than other types of radishes, making them a good choice for spring and fall plantings. They are also less prone to becoming woody or tough, which makes them ideal for fresh eating. In general, European -type varieties of radish are known for their mild, slightly spicy flavour and crisp, white flesh, and are a versatile choice for salads, sandwiches, and garnishes. They are also relatively small in size and have a smooth, round or oblong shape. Some of the most popular European type varieties include:

1. Pusa Himani: This variety was developed at IARI Regional station, Katrain. The roots are pure white, 30 to 35 cm long, crisp, sweet flavoured with mild pungency, semi stumped to tapering with green shoulders. It is suitable for growing throughout the year except winter months in hills. It becomes ready for harvest in 55 to 60 days after sowing and yields about 32 t/ha.	
2. White Icicle: This variety was developed at IARI Regional Station, Katrain. It is an extra early European cultivar, which becomes ready for harvest in 25 to 30 days. Roots are pure white, smooth, thin, medium short, icicle shaped, straight, tapering and bright red with white tip. Roots become pithy within a week after attaining maturity which is a main disadvantage of this cultivar.	
3. Scarlet Globe: It is a quick growing variety. Roots are bright red in colour, round, small with white flesh. It takes 25 to 30 days for harvesting from sowing.	
4. Scarlet Long: It is a table type variety introduced from Europe. Roots are long, tapering with red skin and crispy white flesh. It takes 25 to 30 days to attain maturity after sowing.	
5. French Breakfast: Roots are olive like with red skin and white tips. Flesh is crunchy, mild and sweet in taste.	
6. Rapid Red White Tipped: It is an extra early European cultivar which matures in 25 to 30 days. This cultivar is suitable for spring summer sowing. Roots are small in size, round, smooth, bright red in colour with white tip, flesh is pure white, crispy and snappy and produce short foliage.	

7. Japanese White: This is an introduction from Japan. Roots are snow white, smooth, thick, cylindrical and stumpy with 22 to 25 cm length and 5 cm diameter with pure white, crispy and mild pungent flesh. It matures in 45 days after sowing.	
8. Kashi Sweta: It is a selection made at IIVR, Varanasi from Chetki population. Roots are 25 to 30 cm long, straight and tapering with pointed tip. It takes 30 to 35 days for harvesting from maturity and yields about 45 t/ha.	
9. Pusa Mridula: Suitable for growing in kitchen gardens, container gardens as well as in per-urban areas for higher return. Mature in 28 to 32 days with average yield of 13.5 t/ha. It help to fight against cancer, heart disease and stroke.	

Seed rate and Seed treatment

8 to 10 kg seed of tropical varieties and 10 to 12 kg seed of temperate varieties is sufficient for sowing in a1-hectare area. It has been found that germination is influenced by seed size. The highest germination percentage (81 to 87%) was observed in medium-sized seeds weighing 1.38 g/100 seed in radish cv. *Punjab Safed* (Gill and Singh, 1979).

Hardening of radish seeds by two cycles of wetting and drying improves germination. Seed treatment with GA_3 gives better germination of radish seeds.

Sowing and transplanting

Generally small seeded vegetables respond well to nursery raising and transplanting as they generate numerous fibrous roots after transplanting. In contrast bold seeded vegetables and root vegetables don't respond well to transplanting because once their embryonic roots are damaged, they fail to develop lateral fibrous roots. Because of its tap root system and it is a root crop, the radish seeds are directly sown in the field. Lavanya *et al.* (2014) found that early sowing on 1st October with the closer spacing of 45 cm x 10 cm performed well in respect of yield.

Nutrition Management

Tripathi *et al.* (2017) reported that higher level of Nitrogen fertilizer and spacing maximises the ascorbic acid content and total soluble solids in radish var Kashi Sweta. Khalid *et al.* (2015) reported that the combination of organic, inorganic and biofertilizers has given best result and recommended dosage is NPK @ 80:60:60 kg/ha, FYM @ 20tones/ha and PSB@5kg/ha for radish. MSM Imthiyas and TH Seran (2017) have revealed that 20t/ha compost alone or 10 t/ha compost with half dose of NPK fertilizers as basal application with topdressing is the best choice for increasing the marketable tuber yield of radish. Poudel *et al.* (2018) revealed that 250 kg/ha of Nitrogen was optimum for better yield. Cong *et al.* (2015) reported that B deficiency severely affected the rhythmic stomatal closing and opening, preventing the normal functioning of stomata. Kushwah *et al.* (2020) stated that the application of 75 % NPK + 25 % nitrogen through vermicompost showed highest growth, yield and better return in radish.

Irrigation

Irrigation is an essential factor in the growth and development of radish. Adequate and proper irrigation helps to provide the right amount of moisture to the root system of the radish plant, which is crucial for its growth and health. Inadequate irrigation can lead to wilting, stunted growth, and reduced yields. The frequency of irrigation depends on various factors such as weather conditions, soil type, and the stage of growth of the radish plant. During the early stages of growth, radish plants need to be kept moist but not waterlogged. Over-irrigation can lead to waterlogging, which can cause the roots to rot and result in plant death. In the later stages of growth, the frequency of irrigation can be reduced as the radish plants become more drought tolerant.

When irrigating radish, it's important to avoid wetting the foliage, as this can encourage the growth of disease-causing organisms. Drip irrigation or micro-irrigation systems are effective methods for irrigating radish as they can be used to deliver water directly to the root zone of the plant. This helps to conserve water and reduce the risk of disease. It's also important to ensure that the soil around the radish plants is well-drained, as standing water can also lead to root rot and plant death. If the soil is heavy or clay-like, it may be necessary to incorporate organic matter such as compost or peat moss to improve drainage. In conclusion, proper irrigation is a crucial factor in the growth and development of radish plants. Adequate and appropriate irrigation helps to provide the right amount of moisture to the root system, while avoiding waterlogging and disease. It's important to be mindful of the weather conditions, soil type, and stage of growth when irrigating radish to ensure optimal growth and yields.

Weed Management

Regular weeding should be done to control the seasonal weeds. Generally, two to three weedings are done to keep the field weed free throughout the growing season. Light hoeing can also be done to facilitate the proper root growth.

Intercultural Operations

1. Thinning

Thinning is the removal of excess seedlings which is an important operation to maintain desired plant population. It also reduces competition for light, water and nutrients between seedlings. It is done twice to provide proper conductive conditions. In first thinning the diseased and weak seedlings are removed to a distance of 3 to 4 cm between seedlings and in second thinning excessive seedlings are removed to a distance of 6 to 8 cm within the row.

2. Earthing up

Long rooted radish types often have green shoulders because the spreading roots tend to push through the soil's surface and get exposed to the sun. Additionally, the area that has been exposed to the sun develops a harsh texture, which is not at all a desired character. So, to prevent the root shoulders from excessive heat and greening effect of sunrays proper earthing up is essential. This is done during the early stages of crop when roots start developing. Soil required for covering the root should be taken and placed at the base of the plant. **Sandhu (1986)** reported that one weeding and one earthing is needing during early crop growth for better development of the roots.

Physiological disorders and their management

1. Pithiness

Symptoms: It is characterised by the formation of air spaces in the root, which will eventually leave it spongy, dry, and hollow. Such roots are poor in quality and unfit for consumption.

Causes: Pithiness is governed by genetic factors. It increases markedly with the increases in plant age due to delay in harvesting. It is seen both in field and in storage.

Management: Timely harvest of roots without any delay. Excessive fluctuations in the climate should be avoided. Avoid drought stress or excessively high temperature at the end of the crop growth.

2.Forking

Symptoms: It is a condition where vertical growth of the root is inhibited and results in forking. It is an undesirable character as it deteriorates the quality and market value of the roots.

Causes: It is a feature of clay and hard soil. Presence of unrotten organic matter or any plant refuse causes forking by inhibiting downward growth of roots.

Management: Application of fully decomposed organic matter. Growing of radish in deep, well drained loamy soils. Cultivating resistant varieties like Punjab Safed, Arka Nishanth

3. Splitting

Symptoms: Cracks are caused in the roots of radish which results in splitting.

Causes: Heavy rainfall after a long dry spell. Wide difference in temperature in later stages of root development. Boron deficiency also causes splitting. It is a problem in large rooted varieties

Management: cultivating of resistant varieties like IiHR-1-1, Arka Nishanth

4. Freezing injury

Radish is susceptible to freezing injury. Ryall and Lipton (1972) reported that freezing followed by thawing causes the injured tissue to appear translucent. In severe cases, root softens, loose moisture quickly and shrivel. Parsons and Day (1970) stated that in red radishes the pigment oozes out of the roots with moisture, leaving them yellowish and bleached.

5. Akashin

Symptoms: The growth of the root is checked.

Causes: It is majorly caused due to Boron deficiency. It is also induced by the fluctuation of day and night temperature as well as by low soil moisture.

Management: Borax spray

Major insect pest and their management

1. Aphids: *Myzus persicae*, *Brevicoryne brassicae*, *Lipaphis erysimi*

Aphids are the most destructive type of sucking pest of radish crop. It attacks all the crop plants of Brassicaceae family and survives on the alternate host around the year. It attacks both seedlings and mature crop. They suck sap from the ventral surface of the tender leaves reducing the plant vigour and yield. They excrete honeydew on which sooty mould develops and hinders the photosynthetic activity. It attacks crop at seedling stage, flowering stage and pod formation stage. In case of severe infestation, the plants are completely devitalised, leaves and shoots curl up, become yellow and finally die. Cloudy and humid conditions favour the multiplication of aphid population.

Control

- Remove and destroy the plant portion attacked by aphids.
- Spray the crop with water with pressure.
- Spray the crop with systemic insecticides @ 0.1% Metasystox, Rogor etc

2.Mustard sawfly: *Athalia proxima*

It is one of the common pests of radish and turnip. It attacks crop in both vegetative and flowering stage. The grub feeds on leaves and fruits and make holes by biting them thus damaging the crop. It attacks crop late in the growing season.

Control

- In small areas, hand-picking of larvae is done.
- Dusting the BHC @ 25 to 30 kg/ha is recommended.
- Spray the crop with Carbaryl 0.1% Malathion 0.05% or Dichlorovos 0.05%.

3. Flea beetle: *Phyllotreta* sp.

This may be the most serious pest in some areas. It damages the crop during vegetative stage by biting the leaves and making circular holes in them. It becomes very active in sunny days during the spring months.

Control

- Phyto-sanitary conditions should be adopted.
- Alternate host should be removed from the surroundings.
- Spraying the crop with 0.1% Malathion or 0.4% Sevin at 15 days interval.

4. Root lesion nematode:*Pratylenchus penetrans*

For controlling the population density of root lesion nematode, *Avenastrigosa* was considered to be the most effective antagonist plant.

Major diseases and their management

1. Damping off: *Rhizoctonia solani*, *Pithium* sp. *Fusarium* sp. and *phytophthora* sp.

Fungus attack the crop at seedling stage. In pre-emergence damping off, before the seeds emerge from the soil surface in the early stages of germination, the growth points are killed. In post emergence damping off, due to collar rotting and rapid shrinking of cortical tissue of hypocotyl, the infected seedlings fall down on the ground.

Control

- Long crop rotation should be followed
- Good drainage is required.
- Seed treatment with any copper fungicide like Agrosan GN, Thiram or Foltaf @ 2g/kg of seed or hot water treatment at 50°C for 30 minutes before sowing.
- Soil drenching with 0.1% Captan solution when attack is observed.
- Shallow sowing of seed to give the seed a better aerobic condition.
- Avoid overcrowding of seedlings by manual thinning.

2. Alternaria blight

It is a common disease in rainy season. Fungus affects every part of the plant like leaves, stem, pods and seeds. Slightly raised lesions first appear on leaves, stem, pods and seeds. These lesions enlarge many times as they get older. Concentric rings develop in the necrotic tissue giving a target board like appearance. The disease is seed borne. The infected seeds germinate poorly (Mc Lean, 1947), and hot water treatment or any fungicidal treatment will not improve the germination. Warm, moist, cloudy weather and abundant soil moisture conditions favours the occurrence of this disease.

Control

- Use of disease-free, good quality seed.
- Follow long crop rotation.
- Seed treatment with any copper fungicide like Agrosan GN, Thiram, or Foltaf @ 2g/kg of seed.
- Spray the crop with Blitox 0.4% blitane 0.5% captan 0.3% or Bavistin 0.1% to control the disease.

Yield

Yield varies from cultivar to cultivar. Average yield of Asian cultivars ranges from 15 to 20t/ha and European cultivars ranges from 5 to 7 t/ha.

Harvesting and post-harvest Management

Generally, harvesting is done by lifting the roots up and to facilitate this, light irrigation is given a day before harvesting. Radish reaches maturity 25 to

60 days after sowing depending upon the variety. Early maturing European varieties take 25 to 30 days and Asiatic varieties take 40 to 60 days to reach maturity after sowing. Roots become bitter and pithy if harvesting is delayed. The harvested roots are washed, graded and tied in bundles and transported for marketing. The radish should ideally, be of uniform and similar shape for the variety, well formed, smooth, firm but of tender texture. It should also be free from harvest damage, decay, disease or insects. Bunched radish tops should be fresh in appearance and turgid. Radish cannot be stored at room temperature for more than 2-3 days. However, in cold storage, O topped radish if packed in a polybag with few holes for aeration, can be stored at 0 ^{0}C with 90-95 % relative humidity for about 2 months. Chandra *et al* (2018) found that HDPE film packed roots at 0 degrees Celsius exhibited longer shelf life (more than 2 months).

Modern Technology relevant to vegetable crops

Protected cultivation is one of the modern technologies which creates a favourable environment for the sustained growth even in adverse climatic conditions. This becomes relevant to the farmers of small land holding to produce more crops especially in during the off season when the prices are higher. Protected cultivation has very high entrepreneurial value and profit maximisation leading to local employment. Precision application of water and nutrients is possible through drip fertigation to attain very high crop water and nutrient use efficiency. Hasan *et al* (2019) stated that to reduce the postharvest losses of vegetables, drying of the produce can be done. Drying of vegetables through conventional approaches like sun or open-air drying found a slow process and may lead to the inferior quality product due to contamination. Various advanced drying techniques such as solar, microwave, vacuum, infrared, freeze, oven drying, and different hybrid drying technologies have been developed around the globe as being successfully used for various vegetables.

Rooftop gardening of root vegetables

Rooftop gardening of root vegetables using modern technology is a growing trend in urban areas as people look for ways to grow their own food in limited spaces. With the advent of new and improved tools, materials, and techniques, rooftop gardening has become an increasingly accessible and sustainable option for growing a variety of crops, including root vegetables. One of the key modern technologies used in rooftop gardening is hydroponics, which is the practice of growing plants in water rather than soil. This method not only saves space, but it also eliminates the need for soil, which can be heavy and take up valuable real estate on a rooftop. Additionally, hydroponic systems can be designed to be highly efficient and controlled, ensuring that the plants receive the exact amount of nutrients, water, and light they need to grow.

Another important technology in rooftop gardening is the use of smart systems and automation. For example, there are sensors that can monitor moisture levels in the soil and automatically water the plants when necessary. There are also systems that can track sunlight exposure, temperature, and other environmental factors, and adjust accordingly to provide the optimal growing conditions for the plants. Finally, the use of efficient, compact, and lightweight containers for growing root vegetables is another key aspect of modern rooftop gardening. These containers can be easily transported and set up on a rooftop, allowing even apartment dwellers to get in on the action.

In conclusion, rooftop gardening of root vegetables using modern technology offers a unique solution to the challenges faced by urban farmers, providing a way to grow fresh, healthy food while also reducing the environmental impact of traditional agriculture. With the right tools, techniques, and materials, anyone can start their own rooftop garden of root vegetables and enjoy the benefits of growing their own food in an urban setting.

Future Prospect

Radish is an important crop that is widely cultivated for its edible root, which is high in vitamin C and other nutrients. It is a fast-growing crop that can be harvested in as little as 4-6 weeks, making it an ideal crop for use in intercropping systems, rotation with other crops, or as a catch crop. It is a versatile crop that can be grown in a variety of soil types and climates and is relatively easy to cultivate. In terms of future prospects, crop improvement efforts are focused on developing new varieties of radish that are more resistant to disease, pests, and extreme weather conditions. This can be achieved through traditional breeding methods, such as crossbreeding, or through the use of modern biotechnology techniques, such as genetic engineering. Additionally, research is also being conducted to improve the nutritional content and taste of radish. For example, breeding radish varieties with high levels of antioxidants or other beneficial compounds, or developing radish varieties with a sweeter or milder flavour. Another important future prospect for radish is the use of precision farming techniques. These techniques, such as precision nutrient management and precision irrigation, can help farmers optimize the use of resources and improve crop yields. Precision farming technologies can also help farmers to minimize the use of chemical inputs, such as fertilizers and pesticides, which can help to reduce costs and environmental impact. In addition, radish is also a great crop for biofortification, for example, breeding radish varieties with high levels of micronutrients such as zinc, iron, and vitamin A, to help combat nutrient deficiencies in developing countries. This will be especially important in regions where nutrient deficiencies are prevalent and where traditional food sources may be limited or unavailable. Overall, radish is a valuable crop with many potential uses, and ongoing research and development efforts are likely to continue to improve the crop and its potential in various field. Radish is a versatile crop that can be grown in a variety of soil types and climates, and is relatively easy to cultivate. With the right research and development, radish could become an even more important crop in the future, helping to feed a growing global population and improve the nutritional health of people around the world.

References

Alqasoumi,S., Al-Yahya, M., Al-Howiriny, T.A.W.F.E.Q. and Rafatullah, S.Y.E.D., 2008. Gastroprotective effect of radish``Raphanussativus''L. on experimental gastric ulcer models in rats. *Farmacia-Bucuresti*, 56(2), p.204.

Aykroyd, W.R., Gopalan, C. and Balasubramanian, S.C., 1963. *The nutritive value of Indian foods and the planning of satisfactory diets* (No. 23). New Delhi: Indian Council of Medical Research.

Bose, T.K. and Som, M.G., 1986. Vegetable crops in India. Chadda K L Hand Book of Horticulture (2019). Directorate of Knowledge Management in Agriculture. Krishi Anusandhan Bhavan, New Delhi, 2019,449-453.

Chandra, D., Lee, J.S., Choi, H.J. and Kim, J.G., 2018. Effects of packaging on shelf life and postharvest qualities of radish roots during storage at low temperature for an extended period. *Journal of Food Quality, 2018*.

Cong, X., Jing, H., Lin, N., Xia, Z., Huang, M. and Jiang, X., 2015. Boron deficiency affects cell morphology and structure of young leaves of radish. *Acta Physiologiae Plantarum*, *37*(11), pp.1-10.

Gill, S.S. and Singh, H., 1979. Effect of seed size and sowing dates on germination and yield of radish (Raphanus sativus L.) roots. *Seed Res*, *7*(1), pp.58-62.

Gopalakrishnan T R Vegetable Crops, Horticulture science series-4. Vol-4, New India Publishing Agency, Pitam Pura, New Delhi, 2007, 244-248.

Gopalan, C., Rama Sastri, B.V. and Balasubramanian, S.C., 1971. Nutritive value of Indian foods.

Hasan, M.U., Malik, A.U., Ali, S., Imtiaz, A., Munir, A., Amjad, W. and Anwar, R., 2019. Modern drying techniques in fruits and vegetables to overcome postharvest losses: A review. *Journal of Food Processing and Preservation*, *43*(12), p.e14280.

Hazra, P. and Som, M.G., 1999. *Technology for vegetable production and improvement*. Naya Prokash.

Imthiyas, M.S.M. and Seran, T.H., 2017. Marketable tuber yield of radish (Raphanus sativus L.) as influenced by compost and NPK fertilizers. *Research Journal of Agriculture and Forestry Science ISSN*, *2320*, p.6063.

Jitendra singh. Basic Horticulture. Edn 5, Kalyani Publishers. 2014, 102-114.

Khalid, M., Yadav, B.K. and Yadav, M.P., 2015. Studies on the effect of integrated nutrient management on growth and yield attributes of radish (*Raphanus sativusL.*). *Annals of horticulture*, *8*(1), pp.81-83.

Kirtikar, K.R.B.B. and Basu, B.D., 1935. Indian medicinal plants. Indian Medicinal Plants.

Kushwah, L., Sharma, R.K., Kushwah, S.S. and Singh, O.P., 2020. Influence of organic manures and inorganic fertilizers on growth, yield and profitability of radish (*Raphanus sativus*L.). *Annals of Plant and Soil Research*, *22*(1), pp.14-18.

Lavanya, A.V.N., Sudhavani, V., Reddy, P.S. and Chaitanya, K., 2014. Effect of sowing dates and spacing on growth and root yield of radish cv. pusachetki. *Plant archives*, *14*(1), pp.619-623.

McLean, D.M., 1947. Alternaria Blight and Seed Infection, A Cause. *Journal of Agricultural Research*, *75*, p.71.

Parsons, C.S. and Day, R.H., 1970. *Freezing injury of root crops: beets, carrots, parsnips, radishes, and turnips* (No. 866). US Agricultural Research Service.

Pocasap, P., Weerapreeyakul, N. and Barusrux, S., 2013. Cancer preventive effect of Thai rat-tailed radish (*Raphanus sativus* L. var. caudatus Alef). *Journal of Functional Foods*, *5*(3), pp.1372-1381.

Poudel, P., Shrestha, A. and Shrestha, R.K., 2018. Effect of nitrogen level on growth and yield attributing characters of radish. *Horticult. Int. J*, *2*(4), pp.208-210.

Rana M K Olericulture in India (2008). Kalyani Publishers, pp. 336-363.

Thanburaj S Vegetables, Tubercrops and Spices, Directorate of Knowledge Management in Agriculture. 147-151.

Tripathi, A.K., Ram, R.B., Rout, S., Kumar, A. and Patra, S.S., 2017. Effect of Nitrogen Levels and Spacing on Growth and Yield of Radish (*Raphanus sativus*L.) Cv. Kashi Sweta. *International pure applied bio science*, *5*(4), pp.1951-1960.

13

Advances in Production Technology of Beetroot

Vishal Sharma[1], Puneet Kumar Chhokar[2*], Parmod[3] and Vishnu[4]

[1]*M.Sc (Ag.) Horticulture Vegetable Science Sam Higginbottom University of Agricultural Technology and Sciences, Prayagraj*

[2]*M.Sc. (Ag.) Agronomy Sam Higginbottom University of Agricultural Technology and Sciences, Prayagraj*

[3]*M.Sc (Ag.) Horticulture - Fruit Science Sam Higginbottom University of Agricultural Technology and Sciences, Prayagraj*

[4]*M.Sc (Ag.) Agronomy Sam Higginbottom University of Agricultural Technology and Sciences, Prayagraj*

[*]*Email: chhokarpuneet17@gmail.com*

Introduction

Figure 1.1: Beetroot

Beetroot (*Beta vulgaris L.*) is a very popular herbaceous biennial root vegetable mainly grown for its edible roots and leaves, it is commonly known as Chukander in Hindi, but also has different local names depending on the area, for instance, it is called Bita gacha in Bengali, Bita in Marathi, Beet in Punjab, Salada in Gujarati, Dumpamokka in Telugu, and Bit in Malayalam etc (Kumar, 2015). It belongs to the family Chenopodiaceae (now Amaranthaceae). The chromosome number of beetroot is $2n = 18$. It is used as a cooked vegetable, salad, and for pickling and canning.

Origin and distribution

The origin of beetroot (*Beta vulgaris*) can be traced back to the Mediterranean region and the Middle East. It is believed to have originated from wild sea beet, a plant that grows along the coasts of Europe, North Africa, and Asia. According to Campbell 1979, the beet originated from the wild-beet or sea-beet (*Beta maritrima*) which is a native of Europe. The cultivation of beetroot started thousands of years ago, primarily for its leaves, which were consumed as a green vegetable. Over time, people recognized the edible and medicinal qualities of the beetroot tuber, leading to its domestication and further cultivation. Ancient civilizations, including the Romans and Greeks, cultivated beetroots for their culinary and medicinal purposes. The root's vibrant red color was particularly valued, and it was used as a natural dye.

From its early origins, beetroot cultivation spread to other parts of Europe and gradually reached other continents through trade and exploration. European colonizers introduced beetroot to the Americas, and it became a widely grown crop in North and South America. In modern times, beetroot is cultivated globally in both temperate and tropical regions. Major beetroot-producing countries include Russia, Germany, France, the United States, Poland, and Ukraine. In addition, it is grown in various parts of Asia, Africa, and Australia. Today, beetroot is consumed worldwide for its culinary versatility, nutritional value, and potential health benefits. It continues to be an important crop, and its distribution extends to different continents, reflecting its adaptability to diverse climates and growing conditions.

According to (Hopf *et al.,* 2000) Mesopotamia in the eighth century is when the beet was first mentioned in writing. Later, the Greek traveler Theophrastus compares the beet to the radish, and Aristotle also makes reference to the plant. It was evident from later English and German sources that beetroot cultivation was widespread in medieval Europe (Hill and Langer, 1991).

It is noteworthy that beetroot is different from sugar beet, while most of people think that they are synonyms of one another which is not the case. The easiest way to differentiate sugar beet from beetroot is its white colour whereas beetroot is commonly dark red, though both are from the same family and sugar-beet is used to manufacture sugar and it contains high sugar content but beetroot cannot be used to extract sugar and is more of a table variety (that is for eating purpose). Sugar beet is not extensively grown in India.

Importance and Uses

Beetroot is a versatile vegetable that offers numerous health benefits and finds various uses in culinary and medicinal applications. Its deep red color, earthy flavor, and nutritional richness make it a popular choice in many cuisines worldwide. One of the key reasons for the importance of beetroot is its high nutritional value. It is a good source of dietary fiber, vitamins (such as vitamin C and folate), minerals (including iron and potassium), and antioxidants. These nutrients contribute to overall well-being and support various bodily functions, including immune health, digestion, and cardiovascular health.

Beetroot is commonly used in salads, juices, and smoothies, adding color, flavor, and texture to dishes. It can be consumed raw, roasted, boiled, or pickled, allowing for diverse culinary preparations. In addition to its culinary uses, beetroot has also found application in the food industry as a natural food coloring agent.

Moreover, beetroot has gained attention for its potential health benefits. Research suggests that the antioxidants and betalains present in beetroot may have anti-inflammatory and antioxidant properties, contributing to the prevention of chronic diseases and promoting overall health. It is also believed to support cardiovascular health by helping to maintain healthy blood pressure levels.

Nutritional omposition

Beetroot (*Beta vulgaris*) is a root vegetable known for its vibrant color and unique flavor. It is also packed with essential nutrients. Here is the nutritional composition of beetroot per 100 grams

Table 1: Showing the nutritional composition of beetroot (Sehaj, 2021)

S.No.	Constituents	Amount per 100 g of Beetroot
1	Carbohydrates	9.96 g
2	Fat	0.18 g
3	Protein	1.68 g
4	Water	87.58 g
5	Total Sugar	6.76 g
6	Calcium	16 mg
7	Iron	0.79 mg
8	Potassium	325 mg
9	Sodium	78 mg
10	Vitamin C	3.6 mg
11	Anthocyanin	1.44 - 8.45 mg
12	Carotenoids	1.9 mg
13	Betalins	462 - 838 mg
14	Energy	43 Kcal

Area and Production

Major beetroot-growing states in India are Haryana, Maharashtra, Uttar-Pradesh, and Arunachal Pradesh, while apart from this it is also grown in other states like Punjab, Karnataka, and Tamil Nadu but in small quantities. Beet is grown in an area of approximately 12,500 acres with an annual production of around 900,000 quintals (Natrajan and Veeraragavathatham, 2022).

Botany

As aforementioned beetroot belongs to the root vegetable family having a tap root system which may extend upto a depth of 3m, characterized by thick and fleshy roots that stores a large amount of reserve food while the root colour varies from orange-red to dark-purplish red depending on the traits of the variety under consideration and also due to the presence of β-cyanins, yellow pigment and β-xanthins. At the same, the interior colour may vary depending on the various environmental and other factors like nutrition etc.

Figure 1.2: Beetroot in a field

Leaves are generally ovate with colour varying from dark green to redish. Leaves grows in close spiral forming a rosette with older leaves on the outside and younger ones taking the inner position. The inflorescence of beetroot is large spike which develops in the second year (biennial crop), flowers are small, not easily noticed and are without corolla, but with green calyx. Flowers are composed of 5 stamens, 3 stigma, they are sessile and grows in the bunch of 3-4 on the inflorescence axis and its secondary branches. The true seeds are small, kidney-shaped and brown in colour with girth of around 3 mm.

Taxonomy

Beet is classified into the following categories

Rank	Scientific Taxonomical Details
Kingdom	Plantae
Class	Magnoliopsida
Order	Caryophllales
Family	Chenopodiaceae
Genus	Beta
Species	vulgaris, petula, maritima
Binomial name	Beta vulgaris

Soil and Climate

Beetroot is a cool season vegetable crop but it can also be grown in warm climatic condition, with the optimum temperature for development being 18-21°C, however if a temperature range of 4.5-10°C persists for 15 days or more then it may induce pre-mature bolting.

Loam and sandy loam soil are ideal for the cultivation of beetroot. In addition to these it can also be grown on acidic (pH of <5.8 has detrimental effect on yield) as well as alkaline soil upto a pH of 9-10, while the optimum range being 6-7.

Seed, Sowing and Field Preparation

Ploughing of the field should be done at a depth of 15-20 cm, followed by hoeing to get fine-textured soil, one should ensure that the soil surface is smooth, loose and free from clods. FYM or compost application at the last ploughing is very essential, with the depth of the sowing of seed being 2.5 cm for getting a good stand of the crop.

In India beetroot is sown in march to July in hills, September to November in northern plains and in July to November in southern plains, with the seed rate being 8-10 Kg/ha having a spacing of 45-60cm × 8-10 cm so as to get ideal yield of the crop.

Prior to sowing seeds should be soaked in water to soften the seed coat and getting maximum plant from the sown seed, while their treatment with thiram or captan at the rate of 2.5-3 g/Kg of seed helps in controlling fungal diseases like damping off, downy mildew etc.

Commercial Varieties

On the basis of shape of root beetroot are classified into 5 groups flat, short top, round, half long and long. The following cultivars are recommended for commercial cultivation in India

VarietyName	Description	Picture
Detroit Dark red	Perfectly **round** roots, deep red skin, dark blood red internal colour, heavy yielding cultivar matures in 80 to 100 days.	

Crimson Globe	Roots **globular to flattened**, flesh medium, dark crimson red with indistinct zoning, matures in 90 - 100 days when grown in cool season and 60 - 70 during warm season, large bright green leaves with maroon shade.	
Crosby Egyptian	Roots are **flat globe** with small tap root, internal colour is bright red with some indistinct zoning, has 3-5 leaves with red veins, variety matures in 50-60 days after sowing	
Early Wonder	Roots are **flattened globe with rounded shoulder** and matures in 60-70 days after sowing, has good heat tolerance, suggested for warm season, high yielding variety.	
Ooty 1	Roots are **round**, selection from local type, matures in 120 days.	
Avon Early	Roots are **round to oblong**, matures in 50 - 60 days after sowing, early variety of beetroot, known for its sweet flavor and early maturity	
Ruby queen	Beet is round to oblong shaped, matures in 60 - 70 days after sowing, known for its dep red colour hence used as a natural food colouring ingredient.	

Ochio Canner	Root are **oblonged**, matures in 60 - 70 days after sowing, specifically bred for its ability to be canned and long storage periods,	
Madhur	Known for its high yield and sweet tender flavor.	
Chioggia	A unique variety known for its characteristic red and white rings in its flesh	

Nutrient Management

In beetroot uninterrupted, balanced nutrient supply is required for the production of roots of highest quality, about 25 t/ha of FYM should be applied at the time of last ploughing. While around 70:110:70 Kg/ha of fertilizers N:P:K respectively should be applied to realize full production potential of the crop. Nitrogen is applied in two split doses so as to maximize the nitrogen use efficiency, that is 50% of N along with 100% of P and K can be applied as basal dressing and remaining dose of N should be applied as top dressing 25 days after sowing. In addition to this 20-25 kg/ha of borax is applied for boron deficient soil because beetroot is susceptible to boron deficiency.

Apart from this, it is advisable to conduct soil testing so as to get the detailed description of the nutrients compositions of the soils, in this way deficient nutrients can be identified and accordingly the amendments can be applied and their toxicity or under application can be ameliorated.

Irrigation Management

A steady supply of water is necessary for rapid seed germination and a high quantity of roots. The crop needs 300 mm of water, which can be provided by 5-7 irrigations throughout the rainy season. Excess water delivery should be avoided. While around 75% of the available soil moisture should be maintained, as this crop requires constant moisture supply as mentioned above.

Weed Management

According to Das, 2015 weeds cause around 38% reduction in the crop yield among other forms of losses, in beetroot a variety of winter weeds infests reduction in the growth and yield of beetroot. The weed flora of beetroot constitutes annual grassy weeds, broadleaved weeds and perennial weeds which includes *Polypogon monspeliensis, Anagallis arvensis* and *Cyperus rotundus* respectively.

Following are the 2 approaches that can be undertaken to tackle the menace caused by the diverse weed flora in beetroot

A. **Non-Chemical Approach:** In this approach mechanical and cultural methods are used to control weed population, it includes following methods as mentioned below:

1. **Hand weeding**: It may be defined as the process of manually removing the weeds by hands this is a labour, cost and time intensive method of weed control. Weed control efficiency is high but might be uneconomical as in many instances multiple hand weeding might be needed for effective weed control, generally two hand-weedings at 20 and 40 DAS is enough to get satisfactory weed control.
2. **Earthing up:** Earthing up is typically done when the beetroot is about 2-3 inches tall, and involves gently mounding soil or compost around the base of the stem so that only the leaves are visible. The process can be repeated as the plant grows, keeping the root well-covered and protected. This technique helps to increase the size of the beetroot and maintain its sweet, tender flavor. At the same also helps in supressing the weed growth and giving competitive advantage to the beet by shading the weeds. Apart from controlling weeds earthing-up gives a competitive advantage to the beetroot over weeds.
3. **Interculture:** Refers to the cultural practices performed between the establishment of the crop and its harvest. This includes activities such as soil moisture management, nutrient management etc. These are just a few examples of the cultural practices that can be performed during the inter-culture phase of beetroot production. The specific practices used will depend on the local climate, soil conditions, and other factors. It is important to follow proper cultural practices in order to ensure a successful crop and high-quality beetroot.

B. Chemical Approach: In this approach chemicals are herbicides are used to control weed population, it includes following methods:

1. **Pre-Planting Treatment:** - It is a method of application of herbicide before the sowing of the seeds. For example, Cycloate @ 3 - 4 Kg/ha is effective against annual grasses
2. **Pre-Emergence Treatment:** - It is a method of chemical weed control in which herbicide is applied after sowing but before the emergence of the seedling. For example, Pendimethalin @ 0.75 - 1 Kg/ha is effective against annual grass weed although it also controls broad-leaved weeds.
3. **Post-emergence treatment:** - It is a method of chemical weed control in which herbicide is applied after the emergence of the seedlings or the crop. For example, Diclofop-methyl @ 0.5 - 0.75 Kg/ha is effective against many annual and perennial grass weeds.
4. It is noteworthy that proper spray technology as advised by the experts should be used and under-dosage and over-dosage of the herbicides should be avoided so as to avoid the chances of development of herbicide resistance cases in weeds occupying the field of beetroot.

Physiological Disorders and Their management

1. **Internal Black Spot/ Brown Heart/ Heart rot/ Beet Breakdown:**

   Caused due to boron deficiency in neutral or alkaline soil.

Figure 1.3: Showing Internal Black spot of Beetroot

Symptoms

- Hard and corky black spot are scattered throughout the root
- Boron deficiency first affect young cells and tissues

Management

- Apply boron to the soil in combination with other fertilizers or separately

- Apply 1lb/A actual boron on the leaves twice in mid-season, this also provides good control of boron deficiency

  Note: Deteroid dark red variety of beetroot shows less symptoms of B deficiency

2. **Speckled yellow**: This disorder is caused due to deficiency of "Mn". Mn deficiency is very severe in sandy or very alkaline soil.

Symptoms

- Yellowing is followed by necrotic spot which appear in the clorotic zones
- Leaves with yellow blotches between veins tends to cover in summer

Figure 1.4: Speckled yellow disease of Beetroot

Management

Apply boron to the soil in combination with other fertilizers or separately.

Insect pest and their Management

1. **Beet leaf minor** (*Pegomyia hyocyami*): The female adult lays eggs on the underside of the leaves the affected leaves turn pale yellow get distorted and crumpled, such leaves gradually die away.

Figure 1.5: Beet leaf minor in Beetroot

Management

Spraying the under side of leaf with methyl dematon(0.03%) at the time of emergence of leaves.

2. **Beet leaf hopper** (*Neoaliturus tenellus*): It is the most serious pest of the beetroot because it transmit beet curly top virus that is called curly top disease.

Figure 1.6: Beet leaf hopper in Beetroot

Management

For effective pest control, it is recommended to spray insecticides twice after transplanting. The first application should be done 15 days after transplanting, followed by a second application 30 days after transplanting. You may choose one of the following insecticides for spraying:

- Spraying with dimethoate 0.03% controls the pest effectively.
- Fenitrothion 50 EC: Mix 80 ml of Fenitrothion 50 EC with the appropriate amount of water as per the product instructions.
- Monocrotophos 36 SL: Mix with the appropriate quantity of water following the instructions provided by the product.

Diseases Management

1. **Cercospora leaf spot:** It is a fungal disease caused by *Cercospora beticola.* Symptoms of this disease first appear as individual circular spots that are tan to light brown with reddish-purple borders, the leaves become yellow under highly infected condition.

Figure 1.7: Cercospora Leaf spot in Beetroot

Management

- 3 or more year of crop rotation can reduce the innoculum load
- Growing tolerant variety to cercospora leaf spot
- Seed treatment with captan or thiram @ 2.5 - 3 g/kg seeds
- Spray carbendazim 150 g + mancozeb @ 2 kg/ha

2. **Downy Mildew:** it is also a fungal disease caused by *Peronospora farinosa*

Symptoms: Causes red rimmed spots on the leaves of beet this disease infects the top of beet root causing discoloration, also infect the flowers and seedball

Figure 1.8: Downy Mildew in Beetroot

Management

- Crop rotation
- Eradication of weeds
- Destruction of volunteer crucifer plants
- Spraying ridomil 0.1% or zineb 0.2% at 10 - 15 days interval

3. **Curly top disease**: This disease is caused by beet curly top virus, while the beet leaf hopper is the vector of this disease. Symptoms of this viral disease include warty leaf viens and twisted leaves, plant becomes stunted.

Figure 1.9: Curly top disease in Beetroot

Management

- Early sowing of crop
- Dense population of plants and quick growth tends to decrease the incidence of this disease.
- Three spraying of methyl demeton is recommended

Yield and Storage

- Average yield of beetroot is 25 - 30 t/ha
- Beet storage temperature 0 ^{0}C and 90% Relative Humidity

Harvesting and Post harvest Management

When the beets reach a diameter of 3 to 5 cm, they are picked. Typically, the plants are pulled by hand, the tops are cut off, and the roots are cleaned and sorted by size. Mature roots are mechanically collected, detopped, washed, graded, and put in polythene bags. The shelf-life is extended by decreasing water loss during storage and transportation by removing the tops and packaging the food in polythene. With surface fissures, the large beets became harsh and woody, which is undesirable.

Future Prospect

It is clearly evident that Beetroot is a highly nutritious vegetable that is gaining attention as a potential "superfood" due to its high levels of antioxidants, anti-inflammatory compounds, and nitrates, which can help lower blood pressure and improve athletic performance and at the same time can also reduce the neurodegenerative diseases like Alzheimer's disease etc. (Giampaoli *et al.*, 2023, Szyma *et al.*, 2023 and Delleli *et al.*, 2023). Beetroot is also a good source of vitamins and minerals, including folate, potassium, and iron. In terms of future prospects, beetroot is being researched for its potential health benefits in areas such as cardiovascular health and cancer prevention. Beetroot juice and supplements are becoming increasingly popular as a natural way to enhance athletic performance and endurance, and research suggests that beetroot may also have potential as a treatment for conditions such as hypertension and dementia. Another benefit of beetroot is that it is easy to grow and can be used in a variety of dishes, from salads and soups to juice and desserts, which makes it a versatile ingredient in food industry. Overall, beetroot has a promising future as a functional food and ingredient due to its high nutritional value and potential health benefits.

Crop improvement of beetroot is an important area of research for the future, as it can help to increase yield, improve disease resistance, and enhance nutritional content. One way to improve crop yields is through genetic modification, which can increase the resistance of beetroot to pests and diseases, as well as improve its tolerance to environmental stressors such as drought or extreme temperatures. This can help to increase the overall productivity of beetroot crops and make them more resilient to changing climate conditions. Another area of research is focused on improving the nutritional content of beetroot. Scientists are working to develop varieties of beetroot that have higher levels of beneficial compounds such as antioxidants, anti-inflammatory compounds, and nitrates. This can help to make beetroot an even more valuable food source for preventing chronic diseases such as hypertension and cancer. Another promising area of research is the potential use of beetroot as a biofuel crop. Beetroot is a good source of sugar, which makes it a suitable candidate for biofuel production. Scientists are working on developing new varieties of beetroot that have higher sugar content, which would make it more efficient as a biofuel crop.

References

Campbell GKG (1979) Sugar beet in the evolution of crop plants, (Simmonds, N.W., Ed.). Longmans. N.Y. Green, London, p 130

Chadha, K.L. (2019). Hand Book of Horticulture. Second Edition, Volume 1, Indian Council of Agricultural Research, New Delhi. pp. 480-482.

Das T.K. (2015). Weed Science Basics and Applications. Jain Brothers. New Delhi. Pg: 706 - 708.

Delleli, S.; Ouergui, I.; Messaoudi, H.; Trabelsi, K.; Glenn, J.M.; Ammar, A.(2023) Chtourou, H. Does Beetroot Supplementation Improve Performance in Combat Sports Athletes? A Systematic Review of Randomized Controlled Trials. Nutrients; 15, 398.

Giampaoli O, Ieno C, Sciubba F, Spagnoli M, Miccheli A, Tomassini A, Aureli W, Fattorini L, (2023). Metabolic Biomarkers of Red Beetroot Juice Intake at Rest and after Physical Exercise. Nutrients; 15(9):2026.

Hill G and Langer RHM. (1991). Agricultural plants. Cambridge, UK: Cambridge University Press, 197-199.

Hopf Maria; Zohary Daniel. (2000). Domestication of plants in the old world: the origin and spread of cultivated plants in West Asia, Europe, and the Nile Valley. Oxford [Oxfordshire]: Oxford University Press. 2000, 200.

Kaur. S. (2021) *Utilization of beetroot (beta vulgarisl.) for the development and nutritional evaluation of functional food.* Doctorate thesis, Punjab Agricultural University, Ludhiana, Punjab, India.

Kumar Y.(2015). Beetroot: A Super Food. International journal of Engineering Studies and Technical Approach, 1(3).

Natrajan. S and Veeraragavathatham. D. (2022). Beetroot. Textbook of Vegetable, Tuber crops and Spices. S. Thamburaj and Narender Singh. Directorate of Knowledge Management in Agriculture, ICAR - New Delhi. Pg: 156 - 160.

Selvakumar. R. (2014). A Textbook of Glaustas Olericulture. First edition. New Vishal Publications. New Delhi. Pg: 344 - 347.

Singh .D. (2019). Vegetable Science. First Edition. New Vishal Publications. New Delhi. Pg: 123 - 127.

Swarup. V. (2022). Vegetable Science and Technology in India. Third Edition. Kalyani publishers. Ludhiana. Pg: 556 - 559.

Szyma ´nski, J.; Szwajgier, D.; Baranowska-Wójcik, E (2023). The Role of Beetroot Ingredients in the Prevention of Alzheimer's Disease. *Appl. Sci*, 13, 1044.

14

Advances in Production Technology of Turnip

***Amit Kumar**[*1], **Ramadugu Subhash**[2], **Jagraj Singh**[3], **Satvaan Singh**[4] and **Pradip Kumar**[5]*

[1]*Ph.D. Research Scholar, Department of Fruit Science, Sardar Vallabhbhai Patel University of Agriculture and Technology, Meerut-250110, Uttar Pradesh, India.*

[2]*Ph.D. Research Scholar, Department of Vegetable Science, College of Horticulture-Anatharajupeta Dr. YSR Horticultural University, Andhra Pradesh, India*

[3]*Ph.D. Research Scholar, Department of Vegetable Science, Sardar Vallabhbhai Patel University of Agriculture and Technology, Meerut-250110, Uttar Pradesh, India.*

[4]*Ph.D. Research Scholar, Department of Floriculture and Landscape Architecture, Sardar Vallabhbhai Patel University of Agriculture and Technology, Meerut-250110, Uttar Pradesh, India*

[5]*Ph.D. Research Scholar, Department of Horticulture, Dr. Rajendra Prasad Central Agricultural University Pusa, Samastipur-848-125, Bihar, India*

* *Email: amitworld701@gmail.com*

Introduction

Figure 1.1: Turnip

Turnip (*Brassica rapa* subsp. *repifera* L.) is a root vegetable grown in temperate, subtropical and tropical regions of India that is commonly grown for its edible root and leaves. The root is typically white or cream-colored with a slightly purple or pink top, and has a slightly sweet and earthy flavor. The leaves, which are also edible, have a slightly bitter taste and are typically used in soups, stews, and salads. Turnips are a cool-season crop and are typically planted in the spring or fall. They are hardy and can tolerate frost, making them a popular choice for fall and winter gardens. Turnips are a very nutritious vegetable, they are a good source of vitamin C, fiber, and potassium, as well as other minerals and vitamins. They are also low in calories and high in antioxidants. The root contains a compound called glucosinolate, which has been shown to have anti-cancer properties. The leaves contain a variety of nutrients, including vitamin K, vitamin A, and calcium. Turnips have a long history of cultivation and have been a staple food in many cultures for centuries. They were first domesticated in the Mediterranean and Asia, and have been used in a variety of dishes, such as roasted, mashed, or pickled. In many traditional dishes, they are combined with other root vegetables, such as carrots, potatoes, and parsnips, to create flavorful and hearty meals. Turnips are also a popular crop among farmers, as they are easy to grow, have a short growing season, and are resistant to many pests and diseases. They are also a good crop for small scale and sustainable farming because they can be grown in a variety of soils, require minimal inputs and can be stored for long periods of time. Overall, turnip is a versatile and nutritious vegetable with a long history of cultivation and a wide range of uses in the kitchen. Its versatility in cooking, high nutritional value and easy cultivation make it a valuable addition to any diet and a great crop for farmers. According to the USDA, a 100 g serving of raw turnip greens contains approximately 13,920 IU of vitamin A, which is equivalent to 464% of the recommended daily intake for an adult woman.

According to the USDA, the nutritive value per 100g of the edible portion of raw turnip is as follows:

Table 1: Nutritive value of Turnip fruit/100 g edible portion

Contents	**Amount with Unit**	**Contents**	**Amount with Unit**
Energy	28 kcal	Fatty acids	0.02g
Protein	1.1g	Zinc, Zn	0.2 mg
Total Lipid (fat)	0.1g	Sodium	2 mg
Carbohydrates	6.6g	Potassium	328 mg
Fiber	2g	Thiamin	0.04 mg
Sugars	2.9g	Niacin	0.6 mg
Calcium	44 mg	Vitamin C	28 mg
Iron	0.3 mg	Vitamin B-6	0.1 mg
Magnesium	12 mg	Alpha-tocopherol	0.03 mg
Phosphorus	28 mg	Phylloquinone	2.1 µg

Source: USDA Nutrient Database: kcal = kilocalorie, g = grams, µg = micrograms; mg milligrams

Taxonomy

The wide Brassica family includes turnips (*Brassica rapa* L., 2n = 2x = 20, formerly *B. campestris* subsp. *rapifera*), which have similar yellow (and usually white) blooms. It belongs to a different species called B campestris and does not readily cross with the oleracea (types of cabbage). Turnip and another type of swede, *B. napus* L. var. napobrassica, are closely related. The swede is an amphidiploid vegetable with 38 chromosomes that has been naturally developed from *B. oleracea* (2n = 18) and *B. campestris* (2n = 20). Although neither of the ancestral kinds can be determined with accuracy, it is believed that the cross occurred very often where the two species overlap in their natural environment (from Western Europe to Eastern Asia).

Turnips are classified into the following categories

Rank	**Scientific Taxonomical Details**
Kingdom	Plantae – plantes, Planta, Vegetal, plants
Subkingdom	Viridiplantae – green plants
Superdivision	Embryophyta
Division	Tracheophyta – vascular plants, tracheophytes
Subdivision	Spermatophytina – spermatophytes, seed plants, phanérogames
Class	Magnoliopsida
Superorder	Rosanae
Order	Brassicales
Family	Brassicaceae – mustards, moutardes, crucifers
Genus	*Brassica* L. – mustard
Species	*Brassica rapa* L.

Origin and Distribution

The turnip is a root vegetable that is believed to have originated in the Mediterranean and Asia. It has been cultivated for thousands of years, with evidence of its use dating back to ancient Roman and Greek civilizations. The turnip was traditionally a staple food in many cultures and was often used as a source of food during times of scarcity. The turnip was brought to the Americas by European settlers and quickly became a popular crop in the United States and Canada. It is now widely cultivated throughout the world, with the major producers being China, Russia, the United States, and Japan. It is grown in many temperate regions and can tolerate frost, making it a popular choice for fall and winter gardens. Turnip is also commonly grown in Europe, particularly in countries such as the United Kingdom, Poland, and Germany, where it is used in a variety of dishes, particularly in traditional cuisine. In Africa, the turnip is less common but can be found in some regions such as South Africa and Egypt.

Area and Production

Turnips are grown on 2500 acres of land in India, with a production of 50,000 tonnes per year. It is widely grown in Tamil Nadu, Punjab, Himachal Pradesh, Bihar, and Haryana.

Botany

The turnip (*Brassica rapa* subsp. *rapa*) is a root vegetable that is a member of the mustard family (Brassicaceae). The parts of the turnip plant that are typically consumed include the root and the leaves. The root, also known as the turnip bulb, is the underground storage organ of the plant. It is typically white or pale yellow in color and can be round or elongated in shape. The leaves, also known as turnip greens, are dark green and have a slightly bitter taste. They are edible and can be cooked or used in salads. Additionally, the turnip plant also has a stem, flowers and seed pods. The stem connects the roots and leaves and supports the plant. Flowers are small and white, they develop into seed pods if the plant is allowed to mature.

Soil and Climate

The specific soil and climatic requirements can vary depending on the variety of turnip, The turnip is a cool-season crop that can be grown in a variety of soils and climates, but certain conditions are required for optimal growth and yield. Soil requirements for turnips include well-drained, fertile soils with a pH between 5.5 and 6.5. A soil with a high organic matter content is preferred, as it helps retain moisture and provides necessary nutrients for growth. Sandy soils may require additional fertilization to provide enough nutrients for the turnips to grow well. Heavy clay soils can be amended with organic matter to improve drainage and increase fertility. In terms of climatic requirements, turnips grow best in temperatures between 50-60°F (10-15°C). They can tolerate light frosts and can be planted in early spring or late summer/early fall. Turnips are sensitive to extreme heat and require adequate moisture to grow well. They should be irrigated if the weather is dry or if the soil is dry. Long periods of drought can cause the roots to become tough and woody.

Field Preparation

Field preparation for turnip vegetable cultivation is a crucial step in ensuring optimal growth and yield of the crop. The first step is to conduct a soil test to determine the pH and nutrient levels of the soil. This will help identify any deficiencies that need to be addressed before planting. Once the soil test is done, the field should be cleared of any debris and weeds to ensure a clean and clear seedbed for planting. The field should then be plowed and harrowed to loosen the soil and break up any clumps. This improves soil structure and allows for better water and air penetration. After plowing, the field should be leveled to create a smooth surface for planting. The last step is to incorporate any necessary fertilizers or amendments, such as organic matter, into the soil. This will provide the turnips with the necessary nutrients for healthy growth.

It is important to note that field preparation should be done a few weeks before planting to allow the soil to settle and reach the optimal condition for planting.

Sowing and Spacing

Sowing and spacing are important factors to consider when cultivating turnips. Turnips are typically direct seeded into the field, rather than being started indoors and then transplanted. The sowing time for turnip vegetable cultivation is an important factor to consider for successful growth and yield. In the northern plains of India, Asian-type turnips are sown from July to September whereas European-type turnips are sown from October to December. Sowing season typically lasts from July to September in hills. Its seeds are directly sown along ridges or in lines. The seeds should be sown at a depth of about 1/4 inch to 1/2 inch and should be spaced about 1 inch to 2 inches apart. The ideal spacing depends on the variety of turnips and the method of cultivation. For row planting, a spacing of 2-3 inches between rows and 1-2 inches within the rows is recommended. Plants are arranged in rows that are 10-15 cm apart. Medium seeds don't germinate as well as bold ones perform. 90% to 95% of seeds germinate. In ideal storage conditions, seeds should last four to five years. To germinate, seeds require 4-6 days. For broadcast planting, the seeds can be scattered thinly over the prepared soil and then covered with a thin layer of soil. After sowing, the seeds should be watered well to ensure good seed-to-soil contact and to promote germination. The seedlings should be thinned to the recommended spacing as soon as they are big enough to handle. It is important to note that turnips are relatively fast-maturing crops, so it's best to sow them at the right time to avoid bolting before harvest.

Seed rate

The seed rate for turnips can vary depending on the variety and the growing conditions. In general, a seed rate of around 3 to 4 kg per hactare is recommended for most turnip varieties. However, if the soil is particularly poor or if the weather is dry, a higher seed rate may be needed to ensure adequate plant populations. It's important to note that the seed rate should also be adjusted based on the planting method. For example, if using a drill, a lower seed rate can be used compared to broadcasting the seed. Additionally, it is essential to keep in mind that the seed rate should be adjusted based on the seeding depth and seed size. Larger seeds should be planted at a lower rate compared to smaller seeds.

Important Commercial Varieties

Turnip varieties are divided into two categories: European or temperate and Asian or tropical forms. The Asian or tropical or subtropical turnips are more heat-tolerant, more pungent, and ideal for pickles and its varieties like Punjab Safed 4, Pusa Kanchan, and Pusa Sweti. The European or temperate type turnip is sweeter and tastier and its varieties like Purple Top White Globe, Golden Ball, Snow Ball, Early Milan Red Top, Pusa Chandrima, and Pusa Swarnima. There are many commercial varieties of turnips available, but some of the most popular varieties include:

1. **Purple Top White Globe:** This is a traditional variety that is widely grown for its large, white, round roots and purple tops. It is known for its sweet, mild flavor and is a good option for both spring and fall planting.
2. **Golden Ball:** This is a yellow turnip variety that has a sweet and mild flavor, it is a good option for both spring and fall planting.
3. **Panjab Safed 4:** A cultivar that matures quickly that is frequently planted in Punjab and Haryana. The medium-sized, spherical, and completely white roots have a moderate flavor.
4. **Purple Top White Globe:** It is a large-rooted, heavy-yielding variety. The roots are roughly spherical, with a reddish top half and a creamy lower portion. White, solid, crisp, and with a slightly sweet flavor, the flesh. Small, upright top with cut leaves. It can be grown throughout the cooler months.
5. **Pusa Chandrima:** Its roots have a smooth, pure white skin with fine grains, medium to large size, and virtually flattened globe to globular shape. The meat is soft and delicious. Medium top and lightly cropped leaves It is a heavy cropper with an average yield of 400q/ha and an early maturation period of 50–60 days. In plains, it can be sown from October to December.
6. **Pusa Kanchan:** It is a hybrid of the European (Golden Ball) and Asian (Local Red Round) species. It combines the good characteristics of both parents. The roots match the Local Red Round exactly. The flesh is creamy yellow with red skin and a delicious flavor. Compared to the Local Red Round, the leaf top is shorter. About 10 days after the local parent, it is ready for harvest. Compared to Local Red Round, its roots may be kept in the ground for a longer period of time without becoming spongy.
7. **Pusa Swarnima:** The roots have a flattish spherical shape, creamy-yellow surface, and flesh that is both flavorful and tender. Its medium-sized top has a lightly cut leaf blade. In hills, it can be grown from June to October, and in plains, from October to December. It takes 65 to 70 days for it to reach maturity.

8. **Pusa Sweti:** Pusa Sweti attractive, white roots are ready 45–50 days after planting. It is a relatively early maturing type that can be sown in the plains in October.
9. **Snow Ball:** This cultivar has a modest, medium-sized top and is an early temperate type. It has medium-green, tall, clipped leaves. Round and smooth, the roots have skin that is entirely white. White, fine-grained, fragrant, and sensitive describe the flesh.
10. **Turnip L 1:** 45 to 60 days after sowing, the crop is ready for harvest. It has spherical, white roots that are smooth, crisp, and have a moderate flavour. It produces 105 qtl of roots per acre on average.

Nutrition Management/ Manures and Fertilizers

Nutrient management is an important aspect of turnip vegetable cultivation as it ensures that the plants have a consistent supply of essential nutrients to support growth and development. Proper nutrient management can help to optimize crop yields and improve the overall health of the plants. Manures and fertilizers are essential for turnip vegetable cultivation as they provide the necessary nutrients for the plant to grow. Manures, such as a well-rotted cow or horse manure, can be applied to the soil before planting to improve its fertility and structure. Compost can also be used as an alternative to manure.

The amount of manures and fertilizers that should be applied depends on the soil's texture, pH level, fertility status, and climate. In order to promote the best nutrient uptake by plants for a successful harvest, the appropriate application time is equally crucial. For various agroclimatic zones in India, varied NPK doses have been suggested. However, when preparing the soil, a base dose of 20–25 tonnes/ha of farmyard manure should be applied. Applying 70–100 kg of N and 50 kg/ha of P and K is a complement to this. Before sowing, the full doses of P and K as well as half of the N should be applied. Before sowing, phosphoric and potassic fertilizers are applied 7-8 cm deep. The remaining N is provided in two separate doses, the first during the formation of the roots and the second during the development of the root knobs. The micronutrient requirements are almost identical to those of radish. Turnip has a higher B, Ca, and Mo consumption. In order to remedy the lack of these micronutrients, 1 kg/ha of Micnelf MS-24 is sprayed once or twice, depending on the situation. It helps in turnip management of hollow-stem, pith development, and brown heart.

Water Management

Water management is an important aspect of turnip vegetable cultivation as it plays a crucial role in ensuring healthy growth and development of the plants. Turnips require a moderate amount of water, and the soil should be kept consistently moist but not waterlogged. This is typically irrigated each 8 to 15 days, depending on the weather. An efficient water management system is essential for turnip cultivation, as it ensures that the plants receive the right amount of water at the right time. This can be achieved through a combination of proper irrigation techniques, soil moisture monitoring, and weather forecasting. The most efficient way to provide water to turnips is through a drip irrigation system, which delivers water directly to the roots of the plants, reducing water loss due to evaporation and runoff. This system also allows for precise control of the amount of water applied, so you can ensure that the plants are getting the right amount of water without over-watering.

If a drip irrigation system is not available, overhead irrigation can also be used. However, it is important to avoid wetting the leaves as this can lead to fungal diseases. It is also important to apply water in the morning or evening when the temperature is cooler to prevent evaporation. It is also important to monitor soil moisture levels regularly, as well as the weather forecast, to ensure that the plants are getting the right amount of water. In hot and dry weather, the plants may require more frequent irrigation, whereas in cool and wet weather, the plants may require less irrigation. Another aspect of water management is to ensure that the field is well drained, preventing waterlogging, which can lead to root rot and other issues. Also, it's important to maintain the soil structure, avoiding hard pan layers which can impede the water infiltration. Overall, by implementing effective water management practices, you can ensure that your turnip crop receives the right amount of water at the right time, leading to a bountiful harvest.

Intercultural operations

Interculture operations are the cultural practices that are performed after planting and until harvest in turnip vegetable cultivation. These practices are essential for maintaining and improving crop growth, yield, and quality. Some important intercultural operations include:

Weeding: Turnips should be kept free of weeds to reduce competition for moisture and nutrients. Hand weeding or using shallow cultivation is the best way to control weeds in turnip fields.

Thinning: Turnips should be thinned to the recommended spacing as soon as they are big enough to handle. This will ensure that the plants have enough space to grow and develop properly.

Mulching: Mulching can help to retain moisture, suppress weeds, and keep the soil cool. It is done by spreading organic or inorganic materials around the base of the plant.

Physiological Disorders and their management

There are several physiological disorders that can affect turnip vegetable crops and impact their growth and development. These disorders can be caused by a variety of factors, including environmental conditions and improper cultural practices. Some of the most common physiological disorders of turnip vegetables and their management are:

Bolting: This is a disorder caused by changes in temperature or daylength, which results in the plant producing a flower stem instead of a root. Management includes using varieties with good bolting tolerance and avoiding planting during extreme temperatures.

Nutrient deficiencies: A lack of essential nutrients such as nitrogen, phosphorus and potassium can result in stunted growth, yellowing leaves, and poor root development. Management includes soil testing and applying the appropriate fertilizers.

Major Insect-Pests and their management

Turnip crop is susceptible to several insect pests, including:

Alternaria Leaf Spot: *Alternaria spp.*

The pathogen impacts seeds, pods, leaves, and stems. Small, yellowish, slightly elevated lesions are the first disease symptoms to show on the leaves of seed stems. Later, lesions develop on the seed pods and stems. When it's raining, the infection spreads quickly, and the entire pod may be so diseased that the styler end turns black and shrivels. The fungus enters the tissues of the pod and eventually infects the seeds. The contaminated seed does not germinate. It can be managed by spraying with Mancozeb 0.25%

Aphids: These small, soft-bodied insects suck sap from the leaves and stems of the plant, causing leaf curl and yellowing. They also excrete a sticky substance called honeydew, which can lead to the growth of sooty mold. Management includes using insecticidal soap or neem oil, and introducing natural predators such as ladybugs and lacewings.

Cutworms: These larvae of various moth species cut off seedlings at the soil level, causing significant damage. Management includes using bait traps and rotate the crops to avoid the pests

Flea Beetles: These small, hard-bodied beetles chew small holes in the leaves, causing significant damage to seedlings. Management includes using row covers, and using insecticides such as pyrethrin, neem oil, and rotenone.

Cabbage Loopers: These green caterpillars are the larvae of a moth and feed on the foliage of the plants. Management includes using *Bacillus thuringiensis* (Bt) and neem oil, handpicking the larvae or using traps.

Whiteflies: These small, winged insects suck sap from the leaves and excrete honeydew, causing yellowing and stunted growth. Management includes using yellow sticky traps and horticultural oil.

Preventive measures such as monitoring the field, using row covers, using biological control agents and using chemical pesticides can be used to manage these pests. It is important to use the least toxic approach for the pest management and follow the instructions of the label.

Major Diseases and their management

Turnip crop is susceptible to several diseases, including:

Clubroot: This is a soilborne disease caused by the fungus *Plasmodiophora brassicae.* Symptoms include stunted growth, wilting, and distorted roots. Management includes using resistant varieties, crop rotation, and soil sterilization.

Black Rot: Caused by the fungus *Xanthomonas campestris*, this disease results in blackened leaves and stem, and can lead to complete crop loss. Management includes using resistant varieties, crop rotation, and avoiding overhead irrigation.

Downy Mildew: Caused by the fungus *Peronospora parasitica*, this disease results in yellow spots on the leaves and a white, downy growth on the undersides. Management includes using resistant varieties, crop rotation, and applying fungicides.

Powdery Mildew: Caused by the fungus *Erysiphe polygoni,* this disease results in a powdery white growth on the leaves, which can lead to leaf distortion and reduced yield. Management includes using resistant varieties and applying fungicides.

Root-knot Nematode: These parasitic worms attack the roots of the plant, causing galls and reducing yield. Management includes crop rotation and using resistant varieties.

Preventive measures such as crop rotation, using resistant varieties and using fungicides are the most common ways to manage these diseases.

Maturity Indices

Turnip is a cool-season crop that is typically ready for harvest within 50-70 days after planting, depending on the variety. Maturity indices for turnip include:

Size: Turnips are typically ready for harvest when they reach a size of 2-3 inches in diameter.

Color: Turnips should be a consistent white color with no green or purple areas.

Firmness: Turnips should be firm to the touch, with no signs of softness or decay.

Leaves: Turnip leaves should be dark green and healthy, with no signs of wilting or yellowing.

Roots: Turnip roots should be firm and have a smooth skin without cracks, breaks or pests damage.

It's important to harvest turnips at the right time to ensure maximum quality and yield. Overripe turnips can become tough and woody. Additionally, proper storage conditions are important to maintain the quality of the turnips after harvest.

Harvesting and post-harvest management

The harvest time for turnips depends on the variety and the intended use of the turnips. For turnips grown for their roots, they are typically ready for harvest in about 50-70 days after planting, when the roots are about 2-4 inches in diameter. The harvesting should be done in the evening. For turnips grown for their greens, they can be harvested as soon as the leaves are large enough to use, which is usually about 30 days after planting. When harvesting turnips, it's important to gently pull the roots out of the ground, rather than cutting or breaking them off, to avoid damaging the root. After harvesting, the turnips should be washed, trimmed, and stored in a cool, dark place to keep them fresh.

Post-harvest management of turnip vegetable crops includes steps such as proper harvesting, cleaning, grading, cooling, and storage. Harvesting should be done when the turnips are firm and have reached the desired size. They should be pulled out of the ground carefully to avoid damage. After harvesting, the turnips should be cleaned by removing any dirt or debris. They should then be graded based on size, shape, and quality.

Cooling is an important step as it helps to slow down the respiration rate and prolong the shelf life of the turnips. This can be done by placing the turnips in a cool, well-ventilated area or in a refrigerated storage facility.

The root can be stored safely for 2-3 days under cool and moist conditions. Storage of turnips should be done for 8-16 weeks in a cool (0 ^{0}C or 32-40°F), dark, and humid place, and the humidity should be kept high (90-95%) to prevent the turnips from drying out. Dipping the roots in hot paraffin lowers respiration rate, retards shrinkage, and greatly improves appearance. Proper storage can help to maintain the quality of the turnips for several weeks. It's also important to inspect the stored turnips regularly for any signs of decay or spoilage and to remove any damaged or spoiled turnips to prevent the spread of decay to the rest of the crop.

Yield

The average yield is 20-40 tonnes/ha, which varies with varieties, growing seasons, and management practices.

Future prospects

Turnips are an important vegetable crop with a wide range of nutritional and culinary uses. They are a good source of vitamins and minerals, including Vitamin C, Vitamin K, and potassium. Additionally, they are low in calories and high in fiber, making them a healthy addition to any diet. Turnips are also known to have anti-inflammatory, anti-cancer, anti-oxidant properties and are helpful in maintaining healthy blood pressure. In terms of crop improvement, there are a number of areas where turnips could be improved in the future. One potential area for improvement is in disease resistance, as turnips are susceptible to a number of different diseases, such as blackleg, clubroot, and downy mildew, which can cause significant yield losses. Developing turnips that are resistant to these and other diseases would greatly benefit growers and increase the overall yield and quality of the crop. Another area for improvement could be in increasing the size and yield of the turnips, as larger turnips are often more desirable for both commercial and home growers. This can be achieved through genetic improvement, agronomic practices and better irrigation management.

Additionally, breeding efforts could be focused on developing turnips with more desirable flavor and texture characteristics. As consumer preferences continue to change and evolve, it will be important to develop turnips that are not only nutritious, but also taste great. This could be achieved through conventional breeding or genetic engineering techniques. Furthermore, research and development on developing turnips that are tolerant to abiotic stresses such as drought, salinity, and high temperature will be beneficial for farmers as global warming and climate change will affect the production of turnips.

References

Chadha, K.L. (2019). Hand Book of Horticulture. Second Edition, Volume 1, Indian Council of Agricultural Research, New Delhi. pp. 631-633.

http://indiaplantations.com/index.php?route=product/product&product_id=205

https://www.itis.gov/servlet/SingleRpt/SingleRpt?search_topic=TSN&search_value=23063#null

Nath P. and Swamy K.R.M. (2016). Textbook of Vegetable Crops, Directorate of Knowledge Management in Agriculture, ICAR - New Delhi pp. 254-257.

Ram H. H. (2022) Vegetable Breeding: Principle and Practices. Third Revised Edition 13th Reprinted. Kalyani Publishers Ludhiana-141008. pp. 528-532.

Selsam, Millicent E. and Jerome Wexler. The Carrot and Other Root Vegetables. Morrow Junior Books: New York, 1971.

Selvakumar. R. (2014). A Textbook of Glaustas Olericulture. First edition. New Vishal Publications. New Delhi. pp. 401-402.

Singh D. (2019). Vegetable Science. First Edition. New Vishal Publications. New Delhi. pp. 119-122.

Thamburaj S. and Singh N. (2022). Vegetables, Tuber and Spices Crops, 7th Reprint, Directorate of Knowledge Management in Agriculture, ICAR - New Delhi. pp. 161-164.

USDA Nutrient Database of Turnips on https://fdc.nal.usda.gov/

15

Advances in Production Technology of Pea

***Puneet Kumar Chhokar*[1], *Vishal Sharma*[2], *Gore Subhash Rajendra*[3] *and Naveen Poonia*[4]**

[1] *M.Sc (Ag.) Agronomy, Sam Higginbottom University of Agricultural Technology and Sciences, Prayagraj*

[2] *M.Sc (Ag.) Horticulture - Vegetable Science Sam Higginbottom University of Agricultural Technology and Sciences, Prayagraj*

[3&4] *M.Sc (Ag.) Horticulture-Fruit Science Sam Higginbottom University of Agricultural Technology and Sciences, Prayagraj Correspondence to - Puneet Kumar Chhokar*

Email: Chhokarpuneet17@gmail.com

Introduction

Garden Pea (*Pisum sativum* L.) is a small, round, green seed that is commonly eaten as a vegetable. It is a legume, which means it is part of the same family as beans, lentils, and peanuts. Its chromosome number is 2n=14. Peas are typically eaten fresh, frozen, or canned, and are often used in soups, stews, and casseroles. They are a good source of protein, fiber, various vitamins and minerals. Peas can be grown in a garden or farm and it is considered a staple food in many countries.

Origin and Distribution

Garden peas have a long history of cultivation and have been enjoyed as a food source for centuries. The exact origin of garden peas is believed to be in Central Asia. The distribution of garden peas expanded as they were introduced to different regions and adapted to various climates. They were brought to Europe during ancient times and became popular as a staple crop in many countries. The cultivation of garden peas also spread to Asia, Africa, and the Americas through the process of exploration and colonization. In modern times, garden peas are grown globally in both temperate and cool climates. Major producers of garden peas include China, Russia, India, the United States, and Canada. European countries like France, the United Kingdom, and Germany also have significant production. Garden peas are adaptable and can be cultivated in diverse environments, ranging from the cool, northern regions to warmer climates with adequate water availability. The wild pea plant, known as *Pisum elatius* and this is considered as the ancestor of *Pisum sativum*. Over time, peas were further developed by farmers and selective breeding led to the wide variety of peas we have today. Peas have been an important crop in Europe for centuries, and were brought to the Americas by European settlers.

Importance and Uses

Garden peas (*Pisum sativum*) are highly valued for their nutritional content, taste, and culinary versatility, making them an important vegetable in various cuisines worldwide. These legumes are rich in essential nutrients, including dietary fiber, vitamins (such as vitamin C, vitamin K, and folate), minerals (such as potassium and manganese), and plant-based protein.

The importance of garden peas lies in their contribution to a well-balanced diet. The dietary fiber in peas promotes healthy digestion and can aid in weight management. They also provide antioxidants, such as flavonoids and carotenoids, which have been associated with potential health benefits, including reduced risk of chronic diseases like heart disease and certain cancers.

Garden peas have a range of uses in the culinary world. They can be enjoyed as a standalone vegetable, either raw or cooked, and added to salads, stir-fries, soups, and stews. They are often blanched or steamed to retain their vibrant green color and crisp texture. Peas can also be dried and used in soups, or processed into pea flour or protein isolates for use in various food products.

Nutritional composition

Table 1: Nutritional Composition of Peas (Cervoni, 2022)

S.No.	Constituents	Amount per 100 g of Beetroot
1	Calories	62 kcal
2	Fat	0.2 g
3	Sodium	58 mg
4	Carbohydrates	11 g
5	Fiber	3.6 g
6	Sugar	3.5 g
7	Protein	7.2 g
8	Vitamin K	19.2 mcg
9	Zinc	0.5 mg

Apart from the ones mentioned above it also contains Vitamin C, Vitamin A and other minerals including iron, potassium and zinc. Peas are also low in calories and have virtually no fat, it can be considered as a healthy food option. It's worth noting that some of the nutritional content may vary depending on how the peas are prepared. For example, canned or frozen peas may have slightly lower nutrient levels than fresh peas.

Area and Production

As such pea occupies around 5.9 million ha of the world area with a production of around 11.7 million tonnes, with India contributing around 0.60 million ha to the total world area and around 0.86 million tonnes of production to the total world production (Singh, 2023). Peas are widely cultivated in India and are an important crop in the country. They are grown in many different regions but are particularly prevalent in the northern and western parts of the country. In the northern states of India such as Punjab, Haryana, and Uttar Pradesh, peas are grown as a winter crop. They are also widely grown in the western states of India such as Maharashtra, Gujarat, and Rajasthan. In the eastern parts of India, peas are grown mostly as a summer crop. The states of Odisha, West Bengal, and Bihar are major producers of peas in the eastern region. Peas are also widely grown in the southern states of India such as Tamil Nadu, Andhra Pradesh, and Karnataka. Overall, India is one of the major producers of peas in the world and it's a significant source of income for farmers.

Botany

The pea plant is a member of the legume family having chromosome number 2n = 14, there are many sub-species of pea namely *Pisum arvense* known as field pea that has coloured petals, and *Pisum sativum* which is commonly known as the garden pea, having white flowered horticultural, *Pisum sativum* var. *Macrocarpum* which is edible poded pea, *Pisum sativum* var. *elatius* which is the wild form of pea and *Pisum sativum* var. *Syriacum* which is the wilt form of pea.

The plant is an annual herb that grows to a height of about 2-4 feet. The pea plant has a characteristic compound leaf structure with two leaflets, and a tendril at the end of the petiole. The leaflets are about 1-2 cm long and are arranged alternately on the stem.

The flowers of the pea plant are typically white or purple, and are arranged in clusters at the top of the plant. The flowers have five petals and a characteristic "banner" petal that is larger and showier than the others. The flowers are hermaphrodite, meaning they have both male and female reproductive organs.

The fruit of the pea plant is a pod, which contains several seeds. The pods are typically 2-3 inches long and are green when they are immature and turn yellow when they are ripe. The seeds are smooth and round, and are typically green or yellow in color.

The pea plant is a hardy and versatile crop that can be grown in many different climates and soil types. It is typically planted in the spring, and takes about 60-70 days to reach maturity.

Taxonomy of Pea

Pea is classified into the following categories-

Table 2: Pea Taxonomy

Rank	**Scientific Taxonomical Details**
Kingdom	Plantae
Division	Tracheophyta (vascular plants)
Class	Magnoliopsida
Order	Fabales
Family	Fabaceae (legume family)
Genus	Pisum
Species	sativum

Soil and Climate

Peas are a hardy crop that can grow in a variety of soil and climate conditions, but they do have some specific requirements.

Soil: Peas prefer well-drained, fertile soils with a pH between 6.0 and 7.0. They can tolerate a wide range of soil types, but grow best in sandy loam or loamy soils. Peas also require adequate moisture and good drainage. They also benefit from high levels of organic matter in the soil.

Climate: Peas are a cool-season crop and prefer moderate temperatures. They can tolerate frost and light freezes, but extreme heat can be detrimental to the growth and development of the plant. They grow best at a temperature of 10 - 18°C, with optimum being around 22°C. In general, peas are a versatile crop and can be grown in many different regions. They are commonly grown in temperate regions with cool springs and mild summers, but they can also be grown in subtropical and tropical regions with the right variety and management.

Seed, Sowing and Field Preparation

Seed: Pea seeds are small, round, and typically green or light brown in color. They are usually sold in seed packets or in bulk. When purchasing pea seeds, it is important to check the seed's viability, or ability to germinate, which is usually expressed as a percentage. It's also important to select seeds that are disease-free and well-suited to your growing conditions. The seed rate for the early variety is 100 - 120 Kg/ha and for mid-season and late varieties it is 80 - 90 Kg/ha depending upon the size of the seed.

Sowing: Peas are typically planted in the spring, after the last frost. The seeds should be planted about 1 inch deep in well-drained soil having a pH range between 6.0 and 7.0. before sowing seeds should be soaked in water overnight for better germination, soaking seeds in GA_3 (10 ppm) for 12 hours has given the highest germination.

Field preparation: Before planting, it's important to prepare the field by tilling or turning over the soil to remove any debris and to incorporate any necessary amendments, such as fertilizer or compost, this can be done by plowing the field followed by 1 to 2 harrowings. The soil should also be watered before planting to ensure that it is moist. It is also important to provide support for the pea plants, such as a trellis or a fence, as they will need something to climb on as they grow.

Commercial Varieties

There are several commercial varieties of peas grown in India, some of the popular ones are:

- **Pusa Ageti:** This variety is known for its high yield potential and is tolerant to powdery mildew. It is suitable for both spring and autumn planting.
- **Matar Ageti-7:** This is an early maturing variety released by PAU Ludhiana, which gives picking 60-65 days after sowing and gives a green pod yield of around 90 quintals/hectare (Dhall, 2017).
- **Arka Chaitra:** Tall, tolerant to high temperature, and gives a pod yield of 7 t per ha and has a crop duration of 90 days.
- **Arka Uttam:** Similar to Arka Chaitra except it has long dark green colored pods while Chaitra has light colored pods. Also gives a pod yield of around 7 t per ha and has a crop duration of 90 days.
- **Kashi Samridhi:** Semi-determinant type growth, Resistant to powdery mildew, having dark green pods, and medium maturing duration.
- **Kashi Nandini:** It is an early maturing cultivar of pea developed by pedigree selecting from the cross of P 1542 x VT - 2 - 1 at IVRI, Varanasi. (Anonymous, 2023)

These are some of the common commercial varieties of peas grown in India, but there are other varieties that are popular in different regions of the country.

It's important to select a variety that is well-suited to your growing conditions and local market demand.

Nutrient Management

- Nutrient management is an important aspect of growing peas. Peas require a balanced supply of nutrients for optimal growth and yield. The main nutrients required by peas include:
- Nitrogen (N): Nitrogen is an essential nutrient for vegetative growth and is required for the development of leaves and stems. It should be applied at the rate of 40-60 kg/ha.
- Phosphorus (P): Phosphorus is an essential nutrient for root development, flower and seed production. It should be applied at the rate of 30-40 kg/ha.
- Potassium (K): Potassium is an essential nutrient for the overall health of the plant, including disease resistance and stress tolerance. It should be applied at the rate of 30-40 kg/ha.
- Sulfur (S): Sulfur is an essential nutrient for the growth and development of the plant. It should be applied at the rate of 10-15 kg/ha.
- Calcium (Ca): Calcium is essential for seed formation, root growth and overall plant health.
- Magnesium (Mg): Magnesium is important for the photosynthesis, growth and overall health of the plant.

It is recommended to soil test before applying any fertilizer, and adjust the rates accordingly. Organic fertilizer such as compost, FYM, and vermicompost can also be used as a source of nutrients.

Irrigation Management

Irrigation management is an important aspect of growing peas. Peas require consistent moisture throughout their growing season to ensure optimal growth and yield. The average crop water requirement for field pea was estimated to be 221.0 mm and the average water demand for different crop growth stages of field pea was estimated to be 20.0 mm (initial stage), 52.0 mm (development stage), 100.0 mm (mid-season) and 49.0 mm (late season). The following are some general guidelines for irrigation management of peas:

- Irrigate immediately after planting: Peas should be irrigated immediately after planting to ensure that the soil is moist and to encourage germination.
- Provide consistent moisture: Peas require consistent moisture throughout their growing season. The soil should be kept consistently moist, but not waterlogged.

- Monitor soil moisture: It's important to monitor soil moisture levels and adjust irrigation accordingly. A soil moisture sensor can be used to help determine when irrigation is needed.
- Schedule irrigation: Irrigation can be scheduled based on the crop's water needs and weather conditions. Peas typically need more water during periods of high heat and low rainfall.
- Avoid overhead irrigation: Overhead irrigation can lead to disease and mold, it is recommended to use drip irrigation or furrow irrigation method.
- Irrigation frequency: The frequency of irrigation depends on the weather conditions, soil type and crop stage. In general, it is recommended to irrigate peas once a week during dry weather.

It's important to note that over-irrigation can lead to waterlogged soil and poor root growth, while under-irrigation can lead to stress and reduced yields. Therefore, it is important to monitor soil moisture and make adjustments as needed to ensure proper irrigation management.

Training and Staking

Training and staking are important aspects of growing peas, as they provide support for the plants and help to optimize yield and quality.

Training: Peas are climbing plants and require support as they grow. Training the plants to climb onto a trellis, fence, or other support structure can help to keep the plants upright and prevent them from sprawling on the ground.

Staking: Individual stakes can be used to provide support for each plant. The stake should be inserted into the soil next to the plant and the plant should be tied to the stake as it grows.

Use of netting: A netting can also be used to provide support for the plants. The netting is suspended above the plants and the plants are trained to climb the netting as they grow.

It's important to provide support early in the plant's growth, before they become too heavy and difficult to manage. Supporting the plants also helps to keep them upright, which can improve air circulation and reduce the risk of disease.

It is also important to note that the type of support and method used will depend on the variety of pea, the space available and the personal preference of the grower.

Weed Management

Weed management is an important aspect of growing peas. Weeds can compete with peas for water, nutrients, and light, reducing growth and yield. The following are some strategies for weed management in peas:

- Mechanical control: This can include using tools such as hoes or cultivators to manually remove weeds. This method is most effective when weeds are small and easy to pull.
- Mulching: Mulching with organic materials such as straw or grass clippings can help to suppress weed growth. Mulch also helps to retain moisture and regulate soil temperature.
- Cover cropping: Planting a cover crop before planting peas can help to suppress weed growth and improve soil health. Cover crops such as rye or barley can be tilled into the soil before planting peas, to help suppress weed growth.
- Chemical control: Herbicides can be used to control weeds. However, it is important to select an herbicide that is labeled for use on peas and to follow all label instructions. Pre-emergence herbicides such as pendimethalin @ 0.75 - 1.0 Kg/ha are applied before the emergence of the crop and post-emergence herbicides such as MCPB @ 0.75 to 1.0 Kg/ha at 5 to 6 leaf stage of field pea for controlling broad leaved weeds including *Cirsium arvense* (Canada Thristle).
- Integrated Pest Management: An integrated pest management (IPM) approach combines several methods, such as mechanical control, mulching, cover cropping, and chemical control to effectively manage weeds in a sustainable way.

It's important to note that while chemical control can be effective, it should be used as a last resort, and always consider the environmental impact. It's always best to use multiple methods to manage weeds, such as hand weeding, mulching, and using cover crops.

Physiological Disorder

There are several physiological disorders that can affect peas, some of the most common ones include:

- **Blossom end rot:** This disorder is caused by a calcium deficiency in the plant, it appears as a dark, leathery lesion on the blossom end of the pea pod. It can be managed by ensuring proper soil pH, providing adequate calcium and water, and avoiding fluctuations in soil moisture.

- **Pod distortion:** This disorder is caused by an imbalance in the ratio of potassium to nitrogen in the plant. It can be managed by adjusting the fertilization program to provide the correct ratio of nutrients.
- **Split or hollow pods:** This disorder is caused by uneven watering, which can lead to rapid growth followed by a period of drought. It can be managed by providing consistent moisture to the plants.

Diseases and Their Management

- **Powdery mildew:** This fungal disease caused by Erysiphe pisi leads to the formation of white, powdery spots on the leaves and stems of the plant. It can be managed by providing adequate air circulation, avoiding overhead irrigation, and applying appropriate fungicides like Sulfur, Tabuconazole, etc.
- **Fusarium wilt:** This fungal disease causes the plants to wilt and turn yellow, it can be managed by crop rotation, using disease-free seed and soil solarization.
- **Root rot:** This fungal disease causes the roots of the plants to rot, it can be managed by providing well-draining soil, avoiding over watering and using appropriate fungicides.

It's important to note that it is always best to try to prevent these disorders, by providing the plants with the right growing conditions, such as proper soil pH, adequate water and nutrients, and good air circulation. And to monitor regularly the crop, to detect any symptoms and take action accordingly.

Pest and their management

There are several insect pests that can affect peas, some of the most common ones include:

- **Pea aphid (*Acyrthosiphon pisum*):** This small, green insect feeds on the sap of the plant, causing wilting, yellowing, and distorted growth. It can be managed by using insecticidal soap, neem oil, or insecticides approved for use on peas.
- **Pea weevil (*Bruchus pisorum*):** This small, brown beetle feeds on the leaves, stems, and pods of the plant, causing holes and distortion. It can be managed by using pesticides approved for use on peas and crop rotation.
- **Pea moth (*Cydia nigricana*):** This small, gray moth lays its eggs on the developing pods, causing the pods to become distorted and discolored. It can be managed by using pesticides approved for use on peas, removing infected pods, and crop rotation.

- **Cutworm (*Agrotis ipsilon*):** These small, brown or black caterpillars feed on the stems, leaves and pods of the plant, causing wilting and death of young seedlings. It can be managed by using pesticides approved for use on peas and by creating barriers (such as diatomaceous earth) around the base of the plants to prevent the cutworms from reaching the plants.
- **Thrips (*Kakothrips pisivorus*):** These tiny, slender, winged insects feed on the leaves, flowers, and pods of the plants, causing distorted growth and discoloration. It can be managed by using pesticides approved for use on peas and by keeping the field clean of debris, where they lay their eggs.

It's important to note that it is always best to try to prevent insect pests by providing the plants with the right growing conditions, such as proper fertilization and irrigation, good air circulation, and using crop rotation. Also, monitoring the field regularly for signs of insect infestation and taking action as soon as possible is crucial for pest management.

Yield

The yield of peas in India can vary depending on several factors such as the variety of pea, the growing conditions, and the management practices used. However, on average, the yield of peas in India is around 15-20 quintals per hectare. The yield may also vary depending on the region of cultivation, the soil type and the weather condition. Yields can be increased through proper crop management practices such as selecting the right variety, providing adequate water and nutrients, controlling pests and diseases, and providing proper support for the plants.

It is also important to note that yield may also depend on the season of cultivation, peas grown in the winter season can have higher yield than those grown in the summer season. And using high-yielding varieties, good seed rate and spacing, timely sowing, proper irrigation and fertility management can increase the yield.

Harvesting and Management

Pea harvesting typically occurs when the peas have reached maturity and the pods are dry and brittle. The best time to harvest peas is in the morning when the pods are still cool and the peas inside are fully mature. To harvest, the peas are picked by hand or with a mechanical harvester and then they are dried before being packaged or used for further processing.

Future Prospects

There are a number of future prospects for the pea industry, including:

- Increased demand for plant-based protein: Peas are a good source of plant-based protein, which is becoming increasingly popular as a dietary alternative to meat. This is likely to lead to increased demand for pea-based products such as pea protein powder and pea-based meat alternatives.

- Pea breeding: Pea breeding programs are being developed to create new pea varieties with improved yields, disease resistance, and nutritional content.
- Pea as a sustainable crop: Peas are a relatively low-input crop, making them a sustainable choice for farmers. They are also a good rotation crop, which helps to improve soil health.
- Pea in animal feed: Peas are increasingly being used as an ingredient in animal feed.
- Pea as a bio-based industry: Pea can be used as a feedstock for the production of bio-plastics, bio-fuels, and other bio-based products.
- Pea as a food ingredient: Pea is used in many food products like pea protein isolate, flour, and starch.

Overall, pea has a lot of potential in various industries and research is ongoing to find more ways to utilize this crop.

References

Anonymous (2023). Pea. Indian Institute of Vegetable Research, Varanasi, Assessed on 06-07-2023.

Cervoni B. (2022). Pea Nutrition Facts and Health Benefits. Very well fit. Assessed on 06-07-2023.

Chadha, K.L. (2019). Hand Book of Horticulture. Second Edition, Volume 1, Indian Council of Agricultural Research, New Delhi. pp. 584-587.

Das .T.K. (2015). Weed Science Basics and Applications. Jain Brothers. New Delhi. Pg: 677 - 678.

Dhall, R. K. (2017). Pea Cultivation. Director of Communications, Punjab Agriculture University, Ludhiana.

Prabhakara. B.S, Naik L.B, Mohan. N and Vaaaralakshmi .B. (2022). Textbook of Vegetable, Tuber crops and Spices. S. Thamburaj and Narender Singh. Directorate of Knowledge Management in Agriculture, ICAR - New Delhi. Pg: 196 - 201.

Singh .D. (2019). Vegetable Science. First Edition. New Vishal Publications. New Delhi. Pg: 152 - 159.

Singh. C., Singh P, and Singh. R. (2021). Modern Concepts of Raising Field Crops. Oxford and IBH Publishing Co. Pvt. Ltd., New Delhi

Swarup. V. (2022). Vegetable Science and Technology in India. Third Edition. Kalyani publishers. Ludhiana. Pg: 485 - 497.

16

Advances in Production Technology of Broad Bean

Manish Kumar[1], Mohd Wamiq[2], Amit Kumar[3], Krishna Kaushik[4] and Kushal Chaturvedi[5]

[1&2] Research Scholar, Department of Vegetable Science

[3]Research Scholar, Department of Fruit Science

[4]Research Scholar, Department of Floriculture and Landscape Architecture

[5]Research Scholar, Department of Entomology Sardar Vallabhbhai Patel University of Agriculture and Technology, Meerut-250110, Uttar Pradesh, India

Email: manishkumarpali9311@gmail.com

Introduction

Figure 1.1: Broad bean

The popular legume crop known as broad beans, or *Vicia faba* in scientific jargon, is valued for its diversity and nutritional content. Broad beans have a long history that dates back thousands of years, and they have been used extensively in many cuisines and agricultural practises all over the world. With the use of references from reputable sources, this introduction will discuss the traits, cultivation, and health benefits of broad beans. Annual plants in the Fabaceae family, broad beans have a cool growing season. They originate from the Mediterranean region, but due to their resilience to various climatic conditions, they have been grown in numerous locations across the world. With a height of up to one metre, the plants are renowned for their vigorous growth. Large, podded fruits with many seeds or beans are produced by broad bean plants. Broad bean cultivation is a rather simple process.

Broad beans are a wonderful supplement to any diet from a nutritional perspective. They are a great source of protein, dietary fibre, and important minerals like magnesium, potassium, and iron. Additionally, broad beans are a good source of vitamins, especially folate, thiamine, and vitamin C. Broad beans can help maintain a balanced diet and improve general health when they are included in meals. Broad beans are included into a number of recipes in many culinary cultures around the world. Fresh, boiling, steamed, or stir-fried dishes can all be made using them. Broad beans are frequently used as a side dish, in soups, stews, salads, and other cuisines, demonstrating their adaptability and capacity to improve the flavour and texture of different foods. Young, tender wide beans are sometimes eaten whole, including the pods.

Botany

Their unique botany encompasses different facets of their shape, growth habit, reproductive traits, and genetic diversity. Typically, broad beans grow quickly and have a branching stalk covered in tiny hairs. The leaves are compounded of several leaflets aligned along a central axis, or rachis, and are alternating. Each leaflet is smooth, slightly fleshy, and oval to oblong in form. Large and spectacular, broad bean blossoms are often white or pale pink in colour. They are made up of a keel, two wing petals, and a banner petal, and have an unusual structure that resembles a butterfly. Broad bean plants can pollinate themselves because they have both male and female reproductive organs in the same bloom. The female pistil comprises the ovary, style, and stigma for pollen reception and fertilization, whilst the male reproductive organs, the stamens, create pollen. The process of self-pollination, in which pollen from the stamens is transmitted to the stigma of the same flower, can be used to pollinate broad beans. Cross-pollination can also be caused by outside factors like wind or insects. With numerous cultivars and landraces tailored to various habitats and gastronomic preferences, broad beans demonstrate genetic variety. The inheritance of different features in broad beans has been studied genetically in order to create better cultivars. Inter simple sequence repeats (ISSR) markers, for instance, were used in a study by Naghavi *et al.* (2018) to evaluate the genetic diversity of *Vicia faba*. The broad bean genotypes were discovered to have a lot of variation, which contributed to their adaptability and breeding programme potential.

Taxonomy

The family of legumes and peas known as the Fabaceae includes broad beans as a part. The subfamily Faboideae of this family, which also includes a large number of other legume plants, is where they are categorised. The tribe Vicieae, a group of legumes that also includes several species of vetch, includes broad beans. The broad bean species is known as *Vicia faba*($2n=2x=12$), and the genus Vicia includes a number of other species as well. This taxonomic division offers a framework for researching wide beans' traits, genetics, and relationships to other plant species, which aids in understanding the evolutionary relationships and placement of broad beans within the plant world.

Soil and Climate

Well-drained soils with a high capacity to retain moisture are ideal for broad bean growth. They favour soils with an optimal pH range of 6.0 to 7.0, high organic matter content, and good aeration. For the growth of broad beans, sandy to loamy soils are thought to be optimum. The development of roots and drainage may be hampered by soils with a lot of clay, which might be problematic.

Cool-season plants like temperate areas, such broad beans, which are a crop. Throughout the growing season, they thrive in temperatures between 15°C and 25°C (59°F and 77°F). Although they can withstand a little frost, plants can get harmed by extended exposure to freezing temperatures. It is excellent for broad beans to receive 6 to 8 hours per day with mild sunlight. Though they might produce fewer fruits, they can endure some shade.

Commercial Varieties

Very little or insignificant research has been done in broad bean and such only local and homogenous varieties are available for cultivation. IARI, New Delhi has released some improved variety of broad been.

Pusa Sumeet: Plant is 0.75 m tall having 8-10 branches. Plant bears around 100 pods/plants with average pod length of 6.0cm and thickness of 1.3 cm. The pods of this varieties are attractive, medium long, dark green and borne in cluster. Average pod weight is 1.75 g. The fresh seeds are attractive green and tasty. This is a double purpose broad bean variety. Both tender pods and dried seeds are edible. First picking start after 65-70 days of sowing. it's average yield potential is 18 tones/ha. The variety is suitable for packaging and transport.

Pusa Udit: The pods of this variety are extra-long, flattish and light green. The fresh seeds are attractive green, good in taste. This is a double purpose broad bean variety. Both tender pods and dried pods are edible. It's average yield potential is 30 tones/ha, which is 80% higher than the earlier released variety Pusa Sumeet. The variety is suitable for packaging and transport.

Long Pod Type: Red Epicure.

Winder Type: Imperial Green Windsor, Imperial White Windsor, Giant Four Seeded Green Windsor.

Selection BR-1: Black seeded.

Selection BR-2: Yellow seeded.

Jawahar Vicia 73-31: Dominant type grown in Madhya Pradesh.

Field preparation

The soil for growing broad bean should be suitably prepared to produce the appropriate yield. As a result, the field needs to be loose, friable, deep, and well-drained in order for seeds to thrive. While sufficient soil loosening is necessary for optimal bean development, direct seeded legume crops still need to be deeply and thoroughly ploughed. In advance, prepare the field by repeating 4–5 deep plowings with a disc harrow or cultivator, followed by levelling.

Method of Planting

In the past, fava beans were planted directly in the ground. However, it can also be grown indoors and moved to the main field to ensure early planting.

Direct Sowing

A cool-season crop, fava beans can withstand temperatures of 150 F. Traditionally, fava bean seed is sown directly into the ground as soon as the soil is workable in the spring. It is very advised to plant early to prevent chocolate stain, which is made more likely by the hot weather in July and August.

Transplanting

Early planting is not possible in some years due to the soil and climate. As a result, transplanting may be used as a substitute technique to prevent fava bean planting from being delayed. The fava bean is planted early, which is essential for healthy plants and the highest yield, by sowing the seeds indoors (in a greenhouse) in late March and transferring the seedlings outside. According to our research, seedlings should be grown for 12 to 15 days before being transplanted, and the ideal temperature for growing fava beans in a greenhouse is 600 F. It is not necessary to apply any more fertilizer to seeds that have been properly infected with rhizobia. In a 1.25 inch cell size tray, they planted in pro-mix soil. The recommended depth for planting fava bean seeds is one inch. Planting rows have a 6 inch plant spacing when the row width is 30 inch and a 9 inch plant spacing when the row spacing is 15 inch. Around 28,000–30,000 plants can be found per acre in this planting scheme. In general, 15" rows produce better results than 30" rows by up to 15%, according to our research.

Seed Rate and Plant Spacing

Plant density has an impact on a farming system's canopy growth, radiation interception, dry matter production, evapo-transpiration, weed competition, disease development, plant height, seed yield, and crop production costs. Depending on the cultivar (seed size) and the location, fava bean planting densities fluctuate widely, from 25,000 to 40,000 plants per acre. About 80-100kg seed is required for showing 1 ha Area. Thus, for crop-management purposes, reference is made only to cultivar seed size regardless of botanical type. Set the seed spacing in the beds at 45 x 15 cm.

Irrigation

The *Vicia faba* plant's (broad bean) irrigation needs are influenced by local climatic factors and water resources. Depending on the state and region of the country, different areas may have varying irrigation needs. Generally, broad beans need to be regularly irrigated to keep the soil at an acceptable moisture level during all stages of growth, particularly during flowering and pod development. The day of sowing and once every five days after that see irrigation.

Fertilizers

To increase the yield and quality of broad bean crops, fertilizer treatment is essential. Broad beans have particular nutrient needs, and proper fertilization helps satisfy those needs. The main macronutrients needed for wholesome plant growth are nitrogen, phosphorus, and potassium. Apply 25 t/ha of FYM, 50 kg P, and 25 kg K/ha as a base dose. Between 20 and 25 days after sowing, 25 kg of N and 25 kg of K are applied per ha, and another 25 kg of N is applied between 40 and 45 days.

Weed Management

For resources like sunlight, water, and nutrients, weeds can compete with broad bean crops (*Vicia faba*), resulting in lower yields. To support the successful growth and development of broad beans, effective weed management is essential. Here are some typical methods used to control weeds when growing broad beans:

Before you plant the broad bean crop, get rid of any existing weeds in the field using mechanical techniques like harrowing or cultivation. It also exposes weed seeds to unfavorable environments, which hinders their germination and establishment. Apply organic mulch to the area surrounding the base of the wide bean plants, such as straw or hay. Mulching inhibits the growth of weeds by preventing their seeds from obtaining enough light to germinate. Regularly scan the field for weeds and pull them out by hand when they are little and simple to uproot. During the early growing stages of the crop, hand weeding is extremely crucial. Weeds can be controlled with the use of selective herbicides approved for use on broad beans. To minimize crop damage and maximize weed control, it is crucial to adhere to label instructions and standards for effective herbicide application. When using herbicides, it is essential to thoroughly read and abide by the label's instructions and safety precautions. To effectively eradicate weeds while minimizing any potential harmful effects on the environment and crop health, proper application timing, rates, and techniques are critical.

Major Insect- Pest and their Management

Brown spot caused by bacteria (*Pseudomonas syringae*)

Symptoms

Water-soaked patches on pods that turn brown and necrotic; leaves with small, dark brown necrotic spots that may be encircled by a zone of yellow tissue; pods that twist and distort in the area of infection

Management

Bacterium overwinters in agricultural leftovers; illness is worsened by prolonged wetness of the foliage. Plant only certified seed, rotate your crops frequently, and clear the field of crop detritus after harvest.

Chocolate spot caused by fungus (*Botrytis cinerea)*

Symptoms

There are two types of disease symptoms: aggressive and non-aggressive. The latter is characterized by small, reddish-brown lesions on the plant's leaves, which may also appear on stems and pods. When the disease is aggressive, lesions coalesce and become covered in fluffy mycelium, and large patches of tissue can turn necrotic and die. Long periods of high humidity encourage the transition from the non-aggressive phase to the aggressive phase, which favours overcrowding plants and low potassium and phosphorus levels in the soil.

Management

The goal of control should be to delay the disease's aggressive stage, which can be done by reducing humidity; using the right plant spacing will promote air circulation around plants; Applications of foliar fungicides all season long will shield the crop from chocolate spot.

Downy Mildew Caused By Fungus (*Peronospora Viciae*)

Symptoms

Fungi overwinter in soil and on crop debris, where they can survive for 10 to 15 years and cause yellow-brown blotches on the upper surface of leaves and angular patches of fluffy white-gray fungus on the underside of leaves. Plant growth may also be stunted or distorted and the entire plant may die before flowering.

Management

Crop rotation should last at least five years, crop debris should be thoroughly removed, and late-autumn sowing in deeper soil should be avoided as it can encourage severe illnesses

Fusarium Root Rot Caused By (*Fusarium Salani*)

Symptoms

Brown-red or black streaks on roots that combine as they age; stunted plant growth; yellowing, necrotic basal leaves; lesions that may expand above the soil line. Warm, compacted soils, insufficient soil moisture, and low soil fertility make the disease's damage worse.

Management

Control depends on cultural practices, such as not planting broadleaf crops in the same place more than once every five years and treating seeds with the right fungicide before planting.

Powdery Mildew Caused By Fungus (*Erysiphe Pisi*)

Symptoms

Yellow spots on the tops of the leaves, powdery gray-white patches that combine to blanket the entire plant, and, depending on how badly affected it is, light blue or grey spots on the leaves. Fungus overwinters on plant waste or an alternative host; warm, dry conditions with cool nights that result in dew development favours disease onset.

Management

Plant resistant types, especially if you are sowing them late; employ overhead irrigation (which removes fungus from leaves and lowers viability); sow a crop as soon as possible; To manage severe infestations, Sulphur applications may need to be made often.

Leaf Blight Caused By Bacterium (*Xanthomonas Campestris*)

Symptoms: The spread of illness is highest during humid, rainy weather conditions. Illness can be disseminated by contaminated seed. Bacteria overwinter in agricultural debris.

Management

Plant resistant types exclusively from certified seed, and pre-treat seeds with the proper antibiotic to kill germs before planting. Before symptoms occur, treat plants with a protective copper-based fungicide.

Pests and Their Management

Aphids Caused (*Acyrthosiphon pisum*)

Symptoms

Small soft-bodied insects on the underside of plant leaves and/or stems; usually green or yellow in colour, but may be pink, brown, red, or black depending on species and host plant; if aphid infestation is severe, leaves may yellow and/or become distorted, necrotic spots on leaves, and/or stunted shoots; aphids secrete a sticky, sugary substance called honeydew, which promotes the growth of sooty mould on the plant.

Management

Pruning out the infestation will give control if the aphid population is contained to just a few leaves or branches; scout for aphids on transplants before planting; wherever possible, choose tolerant varieties; Robust plants can be sprayed with a powerful jet of water to knock aphids off leaves; reflective mulches, such as silver coloured plastic, can discourage aphids from feeding on plants; Plants may usually handle low to medium levels of aphid infestation without the need for insecticides; but, very high infestations do require them; The best way of control is typically using insecticidal soaps or oils, like neem or canola oil; Prior to use, carefully read the product labels for any particular usage instructions.

Leafminers (*Lyriomyza spp.*)

Symptoms

Thin, white, twisting leaf trails; excessive mining may cause early leaf drop and white spots on the foliage; Reduced yield may result from early infection; A little black and yellow fly called a leafminer lays its eggs within a leaf; the hatched larvae eat the interior of the leaf. In warm weather, the full lifetime of an insect can be completed in as little as two weeks. An insect may have seven to ten generations per year. Mature larvae drop from leaves into soil to pupate.

Management

Before planting, inspect transplants for leafminer damage; after harvest, remove plants from the ground; and only apply pesticides when leafminer damage has been detected because needless spraying will also lower populations of their natural enemies.

Mexican bean beetle (*Epilachna varivestis*)

Symptoms

Insect damages flowers and small pods as well. The adult insect is an orange-brown beetle with black spots, and the larvae are fat-bodied grubs with tapered ends and rows of noticeable spines. Beetles, which overwinter as adults and reproduce in two to three generations year, can completely destroy bean fields.

Management

Some bean cultivars may make less desirable hosts for beetles, such as snapbeans, which are favoured hosts over lima beans; early types may be less susceptible to beetle damage. Beetle numbers can be decreased by clearing out overwintering areas like bushes and leaves on the ground, handpicking larvae and adults, removing and destroying leaf eggs, and, if an infestation is severe, applying insecticidal soap to the undersides of leaves.

Thrips (*Frankliniella occidentalis*)

Symptoms

During periods of high population, leaves may become deformed, covered in coarse stippling that may make them appear silvery, or flecked with black faeces. a hand lens is optimal for viewing the tiny (1.5 mm), slender insect; Nymphal thrips are smaller and lighter in colour than adult ones, which range from pale yellow to light brown.

Management

Use reflective mulches early in the growing season to stave off thrips; avoid planting near cereals, onions, or garlic because these plants can attract very high numbers of thrips. If thrips become an issue, use the proper insecticide.

Yield

he yield of broad beans differs among different cultivars, but typically an average yield of 10-15 tonnes per hectare of green pods can be achieved.

Harvesting and post-harvest management

Harvesting and post-harvest management of broad beans (also known as fava beans) play a vital role in preserving their quality and maximizing their shelf life. Harvesting should be done when the pods are fully mature and the beans inside have reached their desired size. The pods should be firm and green, indicating optimal maturity. Harvesting can be done by gently pulling the pods from the plant or cutting the stalks. After harvest, it is important to handle the beans with care to avoid damage and bruising. Proper post-harvest management involves removing any damaged or diseased beans and sorting them based on size and quality. The beans should be stored in a cool, dry, and well-ventilated area to prevent moisture build-up and minimize the risk of mold or decay. It is recommended to store broad beans in breathable containers or mesh bags to allow for air circulation. Regular inspection of stored beans is essential to remove any spoiled or sprouting beans promptly. With appropriate harvesting and post-harvest practices, the quality and shelf life of broad beans can be effectively maintained.

Future Prospect

Crop improvement holds promising future prospects for broad beans (fava beans) in terms of enhancing their characteristics and overall performance. Through genetic breeding and biotechnology, broad beans can be improved to exhibit traits such as higher yield, improved disease resistance, tolerance to abiotic stresses like drought or heat, and enhanced nutritional value. Crop improvement programs aim to develop new varieties that are well-adapted to different growing conditions, possess improved agronomic traits, and meet the evolving demands of consumers. Furthermore, efforts can be made to enhance the nutritional composition of broad beans, including increasing their protein content, enriching them with essential vitamins and minerals, and reducing antinutritional factors. By utilizing advanced breeding techniques and genomic tools, researchers can accelerate the process of developing superior broad bean varieties that contribute to sustainable agriculture, food security, and improved human nutrition in the future.

References

CABI Crop Protection Compendium. (2013). *Vicia faba* datasheet. Available at: http://www.cabi.org/cpc/datasheet/56364.

Duke, J. A. (1983*). Vicia faba* L.. Purdue University Center for New Crops and Plant Products. Available at: http://www.hort.purdue.edu/newcrop/default.html.

El-Wakil, W. M., *et al.* (2014). Morphological and anatomical studies on the floral structure of *Vicia faba* L. *African Journal of Agricultural Research*, 9(36), 2757-2762. doi:10.5897/AJAR2014.8882.

FAO. (2019). *Vicia faba* (Broad bean). Retrieved from http://www.fao.org/ag/agp/AGPC/doc/Gbase/data/pf000332.htm.

Garg, A., & Garg, S. (2011). Broad bean (*Vicia faba*): A review. Indian Journal of Natural Products and Resources, 2(4), 439-446.

Kaur, R., *et al.* (2017). Genetics and genomics approaches for improving yield, quality, and drought tolerance in faba bean (*Vicia faba L.*). *Frontiers in Plant Science*, 8, 1877. doi:10.3389/fpls.2017.01877.

Naghavi, M. R., *et al.* (2018). Genetic diversity assessment of *Vicia faba* L. using inter simple sequence repeats (ISSRs) markers. *Journal of Plant Biochemistry and Biotechnology*, 27(3), 314-321. doi:10.1007/s13562-017-0441-9.

Sattell, R., Dick, R. & McGrath, D. (1998). Fava Bean (*Vicia faba L.*). Oregon State University. Available at:http://ir.library.oregonstate.edu/xmlui/bitstream/handle/1957/15226/ em8697.

Schwartz, H. F., Steadman, J. R., Hall, R. & Forster, R. L. (2005) Compendium of Bean Diseases. Available at: http://www.apsnet.org/apsstore/shopapspress/Pages/43275.aspx.

Smith-Spangler, C., *et al.* (2012). Are organic foods safer or healthier than conventional alternatives?: a systematic review. Annals of Internal Medicine, 157(5), 348-366. doi:10.7326/0003-4819-157-5-201209040-00007

17

Advances in Production Technology of Palak

Chhail Bihari[1], Abdul Rahman M[2], Rajat Rajput[3], Amit Kumar[4] and Ajeet Kumar[5]

[1,2,3,&5]Ph.D. Research Scholar, Department of Horticulture, Nagaland University School of Agricultural Sciences, Medziphema, Nagaland

[4]Ph.D Research Scholar, Department of Fruit Science College of Horticulture, Sardar Vallabhbhai Patel University of Agriculture & Technology, Meerut, Uttar Pradesh

Email: rahmanmk1996@gmail.com

Introduction

Palak (*Beta vulgaris L. var. bengalensis)*, also known as spinach beet or perpetual spinach, Saag, Indian Spinach, beet leaf, Swiss chard and Palongpalang. is a leafy green vegetable that is closely related to Swiss chard (*Beta vulgaris* subsp. vulgaris). And it belongs to the family Chenopodiaceae. It is a popular and versatile vegetable known for its nutritious leaves and mild flavour. Palak shares similarities with both spinach (*Spinacia oleracea*) and beetroot (*Beta vulgaris*). The leaves of palak are tender and similar in appearance to spinach leaves. They are dark green and have a slightly glossy texture. The flavour of palak is milder compared to spinach, with a pleasant earthy taste. This makes it a versatile ingredient that can be used in various culinary preparations. Palak is often grown as a cool-season vegetable, as it can tolerate cool temperatures better than spinach. It is appreciated for its ability to produce a continuous harvest of leaves over an extended period, hence the name "perpetual spinach." This characteristic makes it a valuable addition to home gardens and commercial production.

Origin and Distribution

Palak or spinach beet (*Beta vulgaris L. var. bengalensis*) can be traced back to the Mediterranean region. The exact origin of palak is not well-documented, but it is believed to have originated from wild sea beet (*Beta vulgaris* subsp. maritima), which grows along the Indo-Chinese region or coasts of Europe and North Africa. It is a vegetable that has been cultivated for centuries. It has been historically grown in Mediterranean countries, including Greece, Italy, and Spain. From there, its cultivation and popularity spread to other parts of Europe and eventually to other continents. Today, palak is grown in various countries across the globe, both for commercial production and home gardening. It is commonly found in Europe, North America, Asia, and other regions with suitable climatic conditions for its growth. The widespread cultivation of palak is due to its adaptability to different climates and its value as a nutritious leafy green vegetable. Due to its adaptability and popularity, palak or spinach beet has become a widely appreciated vegetable, offering a leafy green alternative to traditional spinach (*Spinacia oleracea*). Its cultivation and distribution continue to contribute to the diversity of leafy greens available in various cuisines worldwide.

Botany

Beta vulgaris is a herbaceous biennial or, in rare cases, perennial plant that can grow to a height of 120 cm (rarely 200 cm); cultivated varieties are often biannual. The roots of the cultivated forms are brown, fibrous, occasionally bloated, and woody in the wild subspecies, or dark red, white, or yellow in the subspecies vulgaris. The stems are simple or branching in the upper section, and their surface is corrugated and striate. In their wild forms, the stems are frequently procumbent. The petioles of the basal leaves are lengthy and may, in certain cultivars, be thickened and red, white, or yellow. The simple leaf blade is whole or undulate, oblanceolate to heart-shaped, dark green to dark red, and slightly meaty. It typically has a noticeable midrib. Inflorescences with dense, spike-like, and basally interrupted flowering are formed. The axils of short bracts or the upper portion of the inflorescence without bracts contain one to three (sometimes eight) flowered glomerules with tiny flowers. These hermaphrodite flowers, which are typically green but may have a reddish hue, are shaped like urns. They are composed of five perianth segments (known as tepals) that are fused at the base, as well as five stamens and a partially inferior ovary containing 2-3 stigmas. The flowers can be pollinated by wind or insects, although wind pollination is more prevalent.

Area and production

Spinach beet is a widely cultivated leafy vegetable found in tropical and subtropical regions. It is commonly grown in states such as Uttar Pradesh, West Bengal, Maharashtra, Rajasthan, Haryana, Punjab, Delhi, Madhya Pradesh, Bihar, and Gujarat, where palak is particularly popular. However, it is important to note that palak does not enjoy the same level of popularity in South India.

Nutritional Values and Medicinal Importance

The plant is rich in minerals, which has led to its nickname "Mines of Minerals." It serves as an affordable and abundant source of iron, calcium, vitamin A, vitamin K, vitamin E, vitamin D, vitamin C, folic acid, thiamine, riboflavin, nicotinic acid, pyridoxine, antioxidants like carotene, flavones, and isothiocyanates, as well as essential amino acids and protein. Along with iron, potassium, manganese, magnesium, copper, and zinc, which are essential for regulating blood pressure and the heart, as well as the antioxidant enzyme superoxide dismutase, which is required to produce red blood cells, the production of sperm, digestion, and the synthesis of nucleic acids. It also has some health benefits, including protection against age-related macular degeneration, nervous system benefits, muscle and bone mineralization benefits, cataract protection, and anti-ulcerative and anti-cancer capabilities. It is generally used as a medicinal herb. It contains a considerable quantity of protein, calcium, and iron, as well as vitamins A and C. The leaves contain very little oxalic acid.

Nutritive Value (per 100 g of edible portion)

Contents with Unit	Amount	Contents with Unit	Amount
Energy (Kcal)	46	Thiamin (mg)	0.26
Moisture (%)	86.4	Riboflavin (mg)	0.56
Protein (g)	3.4	Ascorbic acid (mg)	3.3
Fat (g)	0.8	Ca (mg)	70
Carbohydrate (g)	6.5	Niacin (mg)	380
Vitamin-A (IU)	5862	P (mg)	30

Soil and Climate

Palak is primarily a cool-season crop, but under mild climate's it can be grown all around the year. Compared to other vegetables palak can withstand frost better. Warm weather is not a problem for it, but high temperatures cause early bolting without an acceptable yield. For its cultivation Sandy loam soil having good drainage is ideal. High yields are produced in neutral soil, however, palak can tolerate pH of more than 7. The crop is quite tolerant to saline conditions and can be successfully grown in saline soils. It is sensitive to acidity and liming is required if pH is less than 5.5.

Field Preparation and Spacing

Seeds are immersed in water for the entire night to improve germination. Usually, seeds are sown using a broadcasting technique. But because line sowing makes weeding, hoeing, and harvesting simpler, it is favoured. The crop is sown with 30 cm x 5 cm spacing. Within 10 days of sowing, germination starts. Palak seeds must have enough moisture for germination. For enhanced germination and moisture absorption by the seeds, pre-sowing irrigation is carried out.

Varieties

Palak cultivars can be divided into two groups based on the colour of the midrib and leaf veins.

1. Reddish midrib and leaf veins.
2. Green midrib and leaf veins.

Name Institution	Cultivar	Characteristics
IARI - New Delhi	Pusa Jyoti	• It is a polyploid that developed through selection among All Green colchicine-treated offspring. • It has crisp, succulent leaves that are of the giant-leaved variety. • It produces about 50 tonnes /ha in 6-8 cuttings
	Pusa Bharti	• Green, tender leaves are produced by it. • Compared to Pusa Jyoti, it contains more vitamin C and beta-carotene.
	All green	• This early variety is appropriate for planting in September. • The leaves are uniformly green and tender and have complete leaf margins. • After 75 days, the plant produces seed stalks. • It yields about 12 tonnes/ha
	Pusa palak	• Without any purple pigmentation, it produces uniformly green leaves. • Swiss Chard and Local Palak were hybridized to create • It is a cultivar with late bolting.

IARI Regional Station, Katrain, HP	Pusa Harit	• Suitable for cultivation in the hills throughout the year. • Plants grow upright and vigorously, with consistently thick green, slightly crinkled, giant-sized leaves; they are heavy yielders with an amazing ability to rejuvenate. • It has a late bolting habit and high vitamin A • wide range of adaptability to varying climates and tolerance to alkaline soil. • Average yield: 15-20 t/ha. • Parent- Sugarbeet x Local Palak
PAU-, Ludhiana	Punjab green	• Its plants are semi-erect; the foliage is shining dark green, thick, long, and broad. • The leaves are succulent and free from sourness. • The stem is slightly purple in colour • After 30 days from sowing, it is ready for the first cutting. • 3 tons are produced on average per hectare. • It is a bolting cultivar with a low oxalic acid content.
	Punjab selection	• The leaves are smooth, long, thin, and light green. • They have a mildly sour flavour. Purple colouration covers the stem.
SKNAU- Jobner	Jobner Green:	• originated from a local collection and evolved through natural selection and mutation. • The leaves have a strong flavour and are long, thick, and succulent. • Can be grown effectively in alkaline soils with a pH range of 7.0 to 10.5 degrees. (Tolerant to alkaline soil) • It produces tender leaves and a strong flavour having an entire margin. • The taste is comparable to the cultivar All Green. • Prolific yielder with an average yield of 3t/ ha.

IIHR- Bengaluru.	Arka Anupama	• It was generated through the hybridization of IIHR-10 and IIHR-8, followed by the pedigree selection process. • The leaves are medium-sized, dark green, wrinkled, and appealing. • It is a late bolting variety and regenerates at a fast rate. • 30 days after seeding, the leaves are suitable for harvesting for the first time. • In four cuttings, it yields roughly 4t/ha of fresh green leaves. • Resistant to Cercospora leaf spot.
West Bengal	Banerjee Gaint	• It is the result of a cross between Local Palak and Sugar Beet. • It has huge, thick, succulent leaves, as well as succulent stems and fleshy roots. • It yields 1.5-2t/ha of green leaves.
Maharashtra State Department of Agriculture.	Palak No. 51-16	• It produces multiple cuttings and generates approximately 1.6t/ha. • It is a late bolting cultivar.
Haryana Agricultural University, Hissar	**HS 23**	• Its leaves are dark green, big, thick, soft, and juicy. • It is a quick-growing cultivar and is ready for first cutting 30 days after sowing. • Developed through mass selection
GBPUAT- Pantnagar	Pant Composite-1	• It has a high yield and is resistant to Cercospora leaf spots.
Horticultural Research Station, Ooty of Tamil Nadu Agricultural University	**Ooty 1**	• It is a delicious green leafy vegetable. • It produces 1.5-2 tonnes of leaves per acre. • Vitamins abound in the leaves. • It also has a higher level of carotene. • It can be grown all year and can tolerate frost.

Seed Rate

The single fruit of palak contains 2-3 seeds. It is propagated by seeds. The seed rate varies with the crop seasons. For winter crop 10-15 kg and for summer crop 20-25 kg seed/ha is required.

Sowing Time

In plains, palak is sown in September-November.

In hills, March-May, and August-September.

Manures and Fertilizers

Since it is a leafy vegetable, more nitrogen is required for crop growth. The Standard dosage is 15-20 tonnes FYM and 90:30:30 kg/ha of N, P and K are given. At the time of the last ploughing Full doses of FYM should be applied and incorporated into the soil. As the basal dose applies half of N, the entire P and K should be applied. After every cutting applies the remaining does of N in 3 split doses followed by a light irrigation.

Irrigation

Light irrigation should be given immediately after sowing. Summer crop needs frequent irrigations at 4 to 6 days apart. Irrigations are given at 10 to 12 days intervals during winter and autumn. During the rainy season, crop does not require much irrigation except during long dry spell.

Growth Regulator

The use of GA at 10ml/L in conjunction with 1 % urea has been shown to increase yield in spinach.

Intercultural Operations

Weeding and Hoeing:

At the initial stage of crop plant growth shallow hoeing and regular weeding is essential for better growth. 2-3 hoeing and weeding are required. The soil is kept loose for better growth of the plants. Application of Pre-emergence weedicide like fluchloralin or pendimethalin @ 0.3% is economical to control weeds.

Thinning

Thinning is essential to provide proper plant and leads to get large uniform plants.

Harvesting

Depending on the season and variety, palak leaves are suitable for harvesting 3–4 weeks after sowing. Cuttings between each round are spaced 20–25 days apart. There is a chance of 3–4 cuttings. Just above the roots, the leaves are cut with a sharp knife or sickle. Only trim fragile, succulent green leaves. Compared to summer crop types, winter crop varieties produce more cuttings. Early in the morning, when there is still dew on the leaves, harvesting is not done. After harvesting, the leaves were cleaned, trimmed, graded, and bunched before being sold. Before marketing, throw away any leaves that are sick or yellow.

Yield

Yield varies with cultivar and season. On average, green leaves yield varies from 100-150 q/ha (winter) and 80-100 q/ha (Spring-summer).

Seed production

Similar gardening practices are used for green and seed crops. No more than two or three cuttings should be made for seed production. Seeds are sown with a 60 cm x 20 cm spacing for seed production. It needs greater isolation because it was cross-pollinated. The palak seed crop should be kept apart from fields of other varieties and fields of the same variety that don't comply with varietal purity by 2000 metres for the foundation and 1000 metres for certified seed. The chosen location shouldn't have the same crops as last year. Due to their close kinship, beetroot, swiss chard, and sugar beet can effortlessly cross with palak. Palak must therefore be kept apart from beet, Swiss chard, and other vegetables. Throughout the crop time, seed crop roguing is required to maintain genetic integrity. Minimum 3 inspections are required for roguing, before flowering, at flowering and seed set stage respectively. All off kinds, early bolters, and unhealthy plants should be removed from the field during roguing. The seed must be harvested in about 180 days. When the seeds are fully ripe, the entire plot is harvested. Harvested plants are piled high and let dry for 3 to 5 days on a concrete floor or tarp. Dry plants are beaten with a stick for threshing. After threshing and washing, the moisture content of the seed should be dried to 9% for regular packing and 8% for vapour-tight containers. To store well-dried seeds, use gunny bags with a polyethene lining. Palak's average seed output varies between 8 and 10 q ha-1.

Diseases and pests and Their Management

Damping off:

Causal organism: *Pythium sp.*

Young seedlings are attacked immediately after germination.

Treat the seeds with captan or thiram @3g/ kg of seed and avoid excess moisture.

Cercospora leaf spot

Causal organism: *Cercospora beticola.*

Cercospora leaf spot is a soil-borne disease. Mostly attacked leaf and leaf petioles. Formations of lesions on the edge of the leaf are the first sign of disease incidence, eventually leaving wrinkles, dry and fall. Older leaves are mostly affected. Follow seed treatment practices with carbendazim @ 3g/kg of seed and two to three sprays of copper oxychloride @ 0.3% (3 g/litre of water) or dithane M-45@ 0.2% at 8-10 days intervals is found effective to control the disease.

Aphid

Both nymphs and adults suck the cell sap from leaves, flowers and fruits. Aphid population growth begins at the vegetative stage and peaks at the flowering and seed formation stage. Aphid can be controlled by the application of phosphamidon or metasystox @ 0.05% at 10-15 days intervals.

Leaf-eating caterpillar (*Laphygma exiqua*):

The caterpillar eats leaves for food. It causes considerable damage in the early stage of growth and Yield is reduced. Young larvae scrap the green matter of leaves. Spraying the crop with methyldemeton (0.1%) or phosphamidon (0.1%) or chloropyriphos (0.1%) help to control the attack. If required, repeat spraying at 10-15 days intervals.

Future Prospect

Palak, scientifically known as *Beta vulgaris L. var. bengalensis* holds promising future prospects in terms of crop improvement. Researchers and breeders are actively working towards developing improved varieties of Palak with enhanced traits and characteristics. The focus is on addressing challenges such as disease resistance, adaptability to diverse environmental conditions, increased yield potential, and improved nutritional content. Through genetic studies, selective breeding, and advanced biotechnological approaches, efforts are being made to create Palak varieties that exhibit enhanced resistance against common diseases, tolerance to abiotic stresses, and higher yields. Additionally, there is a growing emphasis on increasing the nutritional value of Palak, including boosting the levels of essential vitamins, minerals, and phytochemicals that contribute to human health. These ongoing crop improvement endeavors aim to ensure the availability of superior Palak varieties, contributing to a sustainable and nutritious food system while meeting the evolving demands of consumers and growers.

References

Ambia, K., Hossain, M., Haque, M., Mosiur, A., Halder, C. and Krisna, S., 2016. Evaluation of seed yield and quality of different genotypes of spinach (*Beta vulgaris*). *Sci. Agri*, 16(2), pp.61-66.

Chadha, K.L. (2019). Hand Book of Horticulture. Second Edition, Volume 1, Indian Council of Agricultural Research, New Delhi. pp. 620-621.

http://eagri.org/eagri50/HORT281/lec31.html

http://ecoursesonline.iasri.res.in/mod/page/view.php?id=12116

Jabeen, A., Narayan, S., Hussain, K., Khan, F.A., Ahmad, A. and Mir, S.A., 2017. Organic production of spinach beet (*Beta vulgaris* var. bengalensis) through the use of manures and biofertilizers. *Journal of Pharmacognosy and Phytochemistry*, 6(6), pp.10-13.

Nayak, P.K. and Maji, S., 2018. Response of nutrient and cutting management for quality and green yield of Palak (*Beta vulgaris* var. bengalensis). *Journal of Crop and Weed*, 14(1), pp.126-129.

Singh, R., Bassi, G. and Sing, D., 2013. Influence of sowing time and leaf cuttings on the seed yield and quality components of palak *(Beta vulgaris* var. bengalensis). *Indian Journal of Agricultural Sciences*, 83(4), pp.410-4.

18

Advances in Production Technology of Spinach

Chhail Bihari[1], *Abdul Rahman M*[2], *Bibarie Mech*[3], *Rajat Rajput*[4] *and Mohd Wamiq*[5]

[1,2,3&4]*Ph.D. Research Scholar, Department of Horticulture, Nagaland University, School of Agricultural Sciences, Medziphema, Nagaland*

[5]*Ph.D. Research Scholar, Department of Vegetable Science, College of Horticulture, Sardar Vallabhbhai Patel University of Agriculture & Technology,*

Email: rahmanmk1996@gmail.com

Introduction

Spinach, scientifically known as *Spinacia oleracea*, is a versatile leafy green vegetable that is widely consumed and enjoyed for its nutritional benefits and culinary uses. Believed to have originated in Persia (modern-day Iran), spinach has been cultivated and consumed for centuries in various regions of the world. It is highly regarded for its vibrant green leaves and is a popular ingredient in salads, soups, smoothies, and cooked dishes. Spinach is an annual plant, meaning it completes its life cycle within one year. It has an upright growth habit with an erect, branched stem that can reach a height of about 30-60 cm.

Due to its delicate flavor and versatility, spinach is used in various culinary preparations. It can be consumed raw in salads or added to sandwiches for a fresh and nutritious crunch. Spinach is commonly cooked as well, either sautéed, steamed, or included in soups, stews, and stir-fries. Its mild taste makes it an adaptable ingredient that complements a wide range of dishes (Thind *et al*., 2021).

Botany

Spinach belongs to the Chenopodiaceae (Goosefoot family) and bearing chromosome is 2n=2x=12. Swiss chard and beets are also members of this vegetable crop family. *Spinacia oleracea* is a fleshy-leaved perennial with a dense rosette of broad, crinkly delicate leaves. Spinach, also known as vilaiti palak, is traditionally categorized as a potherb. It is a popular leaf vegetable growing in temperate climates around the world. It's a dioecious plant with extreme male, vegetative male, and female plants. The extreme male plant is short, whereas the vegetative male and female plants grow taller and produce more fruit. Spinach gets its name from the Persian word 'ispanai,' which means "green hand." There are three forms of spinach: smooth leaf, savoyed leaf, and semi-savoyed leaf. The smooth and semi-savoyed leaves are primarily utilised for processing, whilst the Savoy leaf is primarily used for fresh markets. It is an annual plant with an erect, branched stem that can grow up to 30-60 cm in height. The leaves of spinach are alternate, and simple, and come in various shapes, including triangular, oval, or lanceolate. They have a dark green colour and a smooth or slightly crinkled texture. The small, inconspicuous flowers of spinach are typically green or yellow-green and are arranged in clusters called inflorescences. After flowering, the plant produces hard fruits called achenes, which contain a single seed. Spinach has been cultivated for centuries, resulting in the development of numerous cultivars such as 'Bloomsdale,' 'Tyee,' 'Savoy,' and 'Giant Winter,' which exhibit variations in leaf shape, size, and colour.

Taxonomy

Spinach is classified into the following categories-

Rank	**Scientific Taxonomical Details**
Kingdom:	Plantae
Division:	Magnoliophyta
Class:	Magnoliopsida
Order:	Caryophyllales
Family	Amaranthaceae
Genus:	Spinacia
Species:	Spinacia oleracea

Origin and distribution

Spinach is a native of Central Asia, most likely Persia (Iran), where it has been farmed for more than 2000 years. Spinach was unknown outside of its native region until the early Christian era. It was unknown to the Greeks and Romans at the time. It arrived in China in 647 AD from the surrounding highland kingdom of Nepal. It spread west and was introduced into Spain about 1100 A.D. and Germany in the 13th Century when it quickly became a popular leafy vegetable. In 1828, the first savoyed leaf cultivar was imported into North America.

Nutritional value

One of the notable characteristics of spinach is its rich nutritional profile. It is a great source of vitamins, particularly vitamin A, vitamin C, vitamin K, and folate. It also contains minerals such as iron, calcium, and potassium. Additionally, spinach is low in calories and carbohydrates, making it a popular choice for those seeking nutrient-dense and low-calorie food.

Spinach leaves have a high nutritional content when eaten fresh, steamed, quickly cooked, frozen, or tinned. It's also high in vitamins (A, B_2, B_6, B_9, C, E, K, and folate), minerals (potassium, calcium, magnesium, iron, manganese, and selenium), and fibre. It is also high in active antioxidant chemicals like flavonoids, p-coumaric acid derivatives, and uridine, all of which have anti-oxidative and anti-tumour qualities critical for human health. Vitamins A and C are abundant, as are potassium, folate, and iron. 250 mL (1 cup) of spinach that is cooked contains 42 kilocalories, but the same portion served raw contains 8 kilocalories (Cho *et al.*, 2008).

Importance and Uses

Due to its delicate flavor and versatility, spinach is used in various culinary preparations. It can be consumed raw in salads or added to sandwiches for a fresh and nutritious crunch. Spinach is commonly cooked as well, either sautéed, steamed, or included in soups, stews, and stir-fries. Its mild taste makes it an adaptable ingredient that complements a wide range of dishes. Beyond its culinary appeal, spinach has gained recognition for its potential health benefits. The nutrients present in spinach have been associated with promoting healthy vision, supporting a strong immune system, and maintaining overall well-being. Spinach also contains antioxidants that help protect against cellular damage caused by oxidative stress (Roughani and Miri, 2019).

Varieties

Prickly Seeded: It is a kind with slender, pointed foliage. Plants are erect and robust, with blistering, dense foliage that produces well.

Virginia Savoy: Plants with smooth seeds are upright and strong. Blistered, wrinkled, and thick leaves with spherical tips. The standard yield per acre is 40-50q. It is excellent for the mid and temperate zones.

Banerjee Giant: Plants are tough, with huge, tender leaves that are twice the size of regular spinach and are ideal for the province's subtropical and mid-hill zones. The estimated yield per acre is 60-80q.

Long-Standing: Introduced in the United States. The dark green leaves are big, thick, and conical in shape. 48 days after sowing, the first cutting is possible. The average production per hectare is 100-125 quintals.

Soil Requirements: Spinach (*Spinacia oleracea*) has specific soil and climate requirements for optimal growth and productivity. Here are some key considerations regarding the preferred soil and climate conditions for spinach cultivation:

Soil Type: Spinach thrives in well-drained soil that is rich in organic matter. Loamy soils are ideal for spinach cultivation due to their balanced drainage and water-holding capacity.

Soil pH: Spinach prefers a slightly acidic to neutral soil pH range between 6.0 and 7.0. pH levels outside this range may affect nutrient availability to the plant.

Soil Fertility: Soil should be fertile, with good levels of nitrogen, phosphorus, and potassium. Adequate levels of micronutrients, such as iron, manganese, and boron, are also important for spinach growth.

Climate Requirements:

Temperature: Spinach is a cool-season vegetable that prefers moderate temperatures for optimal growth. The minimum seed germination temperature is 2°C, with a maximum germination temperature of 30°C and an ideal range of 7 to 24°C. Young plants may survive temperatures as low as -9°C. High temperatures can cause spinach to bolt (prematurely produce flower stalks) and negatively affect leaf quality.

Frost Tolerance: Spinach is relatively cold-hardy and can tolerate light frosts. In fact, exposure to cooler temperatures can enhance the sweetness and flavour of the leaves. However, severe frost can damage or kill the plants.

Day Length: Spinach is a long-day plant, meaning it requires longer daylight periods to initiate flowering. This characteristic is important to consider when timing the planting of spinach for different seasons. In terms of cultivation, it is recommended to sow spinach seeds directly in the garden or use transplants. The seeds are typically sown in early spring or late summer to early fall, depending on the climate and desired harvest time. Proper watering practices, such as providing consistent moisture without waterlogging, are essential for spinach growth and preventing common issues like leaf diseases. It is important to note that spinach varieties may exhibit some variations in their specific soil and climate preferences. Therefore, it is always beneficial to consult local agricultural extension offices or gardening resources for region-specific recommendations and guidelines.

Propagation

Spinach is commonly propagated through seeds, and typically, an average of 35 to 40 kg of seeds per hectare is sufficient. However, if the seed lot contains a higher proportion of male seeds, a higher seed rate might be necessary. The germination of spinach seeds can occur within a temperature range of 2-30°C, but the optimum temperature for germination is around 7-24°C. It's important to note that seeds may not germinate well in warm weather conditions. While spinach can tolerate temperatures ranging from 5-24°C, its growth is most vigorous and rapid at temperatures around 15-18°C.

Seed Treatment

Sees should be treated with carbendazim @ 2g per kg of seed. Seed can be precisely sown by sizing it. Fresh seed germinates rapidly at soil temperatures as low as 3 to 4°C, with great outcomes around 10 to 16°C. Increasing temperatures result in faster emergence but a lower percentage of germination. More than a year-old spinach seed rarely germinates at more than 80%. Older seed is even little viable, germinating slowly and unevenly.

Seeding/Planting

When utilizing non-precision seeders, sow at a rate of 35 to 40 kg of seed per hectare. If the crop is precision seeded, less than one-third of this will be required. Sow seed at a depth of 0.5 to 1 cm, with 25 seeds per meter of row for fresh market products and up to 40 seeds per meter of row for processing production. Depending on the cultivation equipment and plant density, rows can be set 18 to 60 cm apart. Where weed management is not an issue, seed can be disseminated. Plant seeds outdoors as soon as the ground can be worked in the early spring through mid-to-late May, and then again in August for crops that will mature in the fall. Seed in early September for field overwintering. From mid-October to mid-November, seeds can be sown in tunnel homes for protected crop overwintering. Winter security can be provided by a floating row cover.

Sowing

Spinach is grown in the Atlantic region from April to May for seeding, with maturity from June 1st to mid-July. In addition, the crop is seeded in August and early September (up to 30-35 days before the first autumn frost) for harvest in September and October. If given protection and/or favorable winter conditions (a snow cover), this crop can also be overwintered.

Sowing time

Zone	Sowing Time
Sub-Tropical	August to December
Intermediate (Mid)	June to September
Intermediate (High)	March-June

Spinach is seeded on flat soil or in raised beds of various sizes, either scattered or in rows. Spinach seed has a short shelf life and should be planted as soon as possible. The seed rate ranges between 12 and 15 kilograms per acre. The seed rate in spinach is high because half of the plants are males with weak growth that cannot be removed before the seed stalk forms. Sowing spinach is similar to sowing beetroot. Plant x row spacing is maintained at 30cm x 5-7.5cm when sown in lines. Plant spacing inside the row is maintained via thinning.

Manures and Fertilizers

Spinach grows quite quickly, especially in the last three weeks before harvesting, when it gains 60% of its fresh weight and takes 60% of its nutrients from the soil. During this time, the primary qualities of colour and succulence are highly dependent on nutrients (Ali *et al.*, 2013). Spinach and spinach beetroot have similar nutrient requirements. However, for cultivars produced for processing, approximately double the amount of nitrogen fertilizer is applied (El-Saady, 2016).

Irrigation

Spinach seed requires adequate soil moisture to germinate. If there is insufficient moisture in the soil, it is critical to provide mild irrigation shortly after sowing. To guarantee consistent and maximum germination, the second watering is administered soon after the first. Because spinach is a shallow-rooted crop, subsequent irrigations are done every 10-15 days. Ample soil moisture is vital during harvesting since it boosts production and quality. To reduce the occurrence of foliar diseases, avoid using the above sprinklers.

Weed Management

Weed control is critical because spinach plants can't compete with weeds. In addition, harvesting becomes harder in weed-infested crops. Weeds can be controlled with two to three shallow cultivations.

Harvest

Spinach is ready to harvest 40-50 days after planting. Cuttings at three-to-four-week intervals are conceivable. Harvesting can be controlled to meet specific needs, ranging from the 5-6 leaf stage through right before seed stalk production. A plant with a seed stem is thought to be unmarketable. The spinach is chopped just above the root. It is then washed, graded, bunched, or bagged. Harvesting of spinach typically begins in the first week of June and continues until mid-July, or until weather and cultivar dictate otherwise. Depending on the local environment (fall frosts), the fall harvest might begin in early September and last until late October. In general, cultivars with more savoyed leaves were utilised for fresh markets, while those with smoother leaves were employed for processing. However, because they are easier to wash and show a quality product, smoother-leaved cultivars are currently used more in the packaged spinach industry. Currently, people frequently use this product as a salad vegetable. April and May could see earlier spring production because of tunnel houses. Most of the crop is grown for roadside markets or is sent directly to wholesalers, making spinach a minor crop at the moment. There may be room for additional wholesale markets given that the commodity has gained popularity as a salad item. Additionally, it might be used as a crop for processing. Processing spinach is harvested above the growing point, leaving only the leaves, and allowing the plant to grow for future harvesting.

Diseases Management

Downy Mildew (fungus)

Characteristics: Indefinite yellowish patches appear on the upper leaf surface. A grey mould appears on the lower surface. Affected areas blacken and perish. Disease thrives in chilly, wet conditions.

Control: To deal with this illness, utilize disease-resistant cultivars. Use three-year crop rotations and treat the seed with hot water.

Fusarium Wilt and Root Rot (fungus)

Characteristics: Plants start to discolour and wilt with older foliage. The fungus damages feeder roots first, followed by tap roots.

Control: Plant the spinach when soil temperatures are high (from mid-June to late August). Tolerant cultivars include “Packer.”

Spinach Blight or Yellows (virus)

Characteristics: Cucumber mosaic virus causes the disease, which is spread by aphids. Early inner leaves get mottled, then yellow, and are eventually dead. Older leaves have become yellow gradually. At higher temperatures, symptoms appear significantly quickly.

Control: Use resistant cultivars.

Insects Pest Management

Spinach Leaf Miner

Characteristics: The adult is a 4 mm-long, slender gray insect. They are usually observed hovering over host plants. Blotches appear because tiny green or white maggots feed inside the leaves. Each year, there are several generations, and the pupae spend the winter underground. The undersides of leaves are where flies lay their eggs. Larvae eat inside the leaves before making their way down to the ground to pupate before becoming flies. The larvae mine between the surfaces of the leaves, resulting in colourless blotches with a silvery overcast. As a result, the produce is unmarketable. Infestations have a negative impact on growth.

Control: When mining damage is first noticed, insecticides should be used. It is possible that two to three treatments at 10-day intervals will be required. Infestations can be reduced early in the season by eliminating infested crop remnants and weeds. Leaf miner has not been a concern in major commercial fields.

Aphids Characteristics: Aphids are little, soft-bodied insects that move slowly. They frequently form enormous colonies on the underside of leaves. A colony is made up of winged and wingless adults, as well as nymphs of varied sizes. Aphids can be black, yellow, or pink, although they are typically green. Aphids obtain their food by sucking plant sap. Saliva injected during feeding could include plant viruses or be hazardous to the host plant. Large numbers of feeding discolour the foliage and twist the leaves. The plants may be covered by honeydew, a sticky fluid secreted by aphids.

Control: Apply only when aphids are so abundant that they cause leaf withering in dry conditions. Ladybird beetles (orange with black stripes or spots) and their larvae (black or purple with orange or white markings) frequently suppress severe aphid attacks.

Yields

Spinach yield varies from 100 to 150 quintals per hectare with respect to cultivar and growing circumstances. Processing varieties produce between 200 and 250 quintals per hectare.

Post-Harvest Handling

Spinach has a rapid respiration rate and a huge leaf surface-to-weight ratio. To avoid wilting and weight loss, quickly and rapidly cool the harvested produce. In a few minutes, vacuum cooling may cool spinach from 19.4 to 28°C. suggested transport and storage temperatures are 0°C and 90% relative humidity. Spinach can be stored in the above-mentioned conditions for 10-14 days. Even when refrigerated at 4°C for eight days, it loses most of its folic acid and carotenoid content. Spinach must be fresh frozen, cooked and frozen, or canned for longer storage.

Seed Production

Seed production of spinach is under-taken in areas having a long cool growing season and is grown the same way as the fresh vegetable crop. It is a highly cross-pollinated crop due to dioecious sex form and a minimum of 1000m isolation distance between two varieties is required to produce genetically pure seed. The seed crop is monitored at least twice to rogue-out off-type and diseased plants, once during the vegetative growth stage and the second during flowering. It is prone to shattering and the crop is harvested when partially dry. After drying for three to four days, it is then threshed, seed cleaned and dried to nine percent moisture or less and stored. Average seed yield is four to five quintals per hectare.

Future Prospect

The future prospects for spinach (*Spinacia oleracea*) in terms of crop improvement are promising. Researchers and breeders are actively working to develop improved spinach varieties with enhanced traits and characteristics. These advancements aim to address challenges such as disease resistance, environmental adaptability, yield improvement, and nutritional enhancement. Through selective breeding and genetic engineering techniques, efforts are being made to create spinach varieties that are more resistant to common diseases, adaptable to various climatic conditions, and capable of higher yields. Additionally, there is a focus on increasing the nutritional value of spinach, including boosting the content of essential vitamins, minerals, and beneficial compounds. These ongoing crop improvement endeavors hold great potential for ensuring the availability of improved spinach varieties that meet the evolving demands of growers and consumers, contributing to a sustainable and healthy food system.

References

Ali, A. H., Hafez, M. M., Mahmoud, A. R. and Shafeek, M. R. 2013. Effect of Bio and chemical fertilizers on growth, yield and chemical properties of spinach plant (*Spinacia oleracea* L.). *Middle East J. Agric. Res.*, **2** (1): 16-20.

Chadha, K.L. (2019). Hand Book of Horticulture. Second Edition, Volume 1, Indian Council of Agricultural Research, New Delhi. pp. 617-619.

Cho, M. J., Howard, L. R., Prior, R. L. and T. Morelock, 2008. Flavonoid content and antioxidant capacity of spinach genotypes determined by high-performance liquid chromatography/mass spectrometry. *J. Sci. Food Agric.*, **88**:1099–1106.

El-Saady, W.A. 2016. Spinach (*Spinacia oleracea* L.) growth, yield and quality response to the application of mineral NPK fertilizer ratios and levels. *Middle East J*, **5**(4), pp.908-917.

Roughani, A. and Miri, S. M. 2019. Spinach: An important green leafy vegetable and medicinal herb. In *The 2nd International Conference on Medicinal Plants, Organic Farming, Natural and Pharmaceutical Ingredients* (pp. 1-6).

Thind, S. K.and Mahal, J. S. 2021. Package of practices for cultivation of vegetables, Additional Director of Communication for Punjab Agricultural University, Ludhiana.

19

Advances in Production Technology of Lettuce

Mohd Wamiq[1]*, Amit Kumar[2], Ramadugu Subash[3] and Chhail Bihari[4]

[1] *Ph.D Research Scholar, Department of Vegetable Science College of Horticulture, Sardar Vallabhbhai Patel University of Agriculture & Technology, Meerut*

[2] *Ph.D Research Scholar, Department of Fruit Science College of Horticulture, Sardar Vallabhbhai Patel University of Agriculture & Technology, Meerut*

[3] *Ph.D Research Scholar, Dr.YSRHU-College of Horticulture, Anantharajupeta*

[4] *Ph.D. Research Scholar, Department of Horticulture, Nagaland University, School of Agricultural Sciences, Medziphema, Nagaland*

* *Email: mohdwamiq0704@gmail.com*

Introduction

Figure 1.1: Lettuce

The Asteraceae family includes *Lactuca sativa*, better known as lettuce, which is cultivated as a leafy vegetable. Due to its usage in the creation of salad, soup, and vegetable curries, this plant is well-known around the world. Additionally, this herb has fantastic therapeutic qualities. Because of its flavor and excellent nutritional content, lettuce is one of the most popular raw green leafy vegetables and is grown all over the world. It is thought to be a crucial source of phytonutrients. Significant physical and genetic differences define it. The Egyptians were the first people to plant this leafy crop. It is currently manufactured all over the world and used in forensic medicine to treat a variety of conditions, including as pain, inflammation and issues with the stomach, as well as urinary tract infections. Its pharmacological potential, which includes antibacterial, antioxidant, neuroprotective, and hypnotic actions, has been scientifically supported by a number of research. The plant's chemical makeup showed that many types of secondary metabolites, including terpenoids, flavonoids, and phenols, were present and should be in charge of the biological activities of the plant. Additionally, the plant has vital components including vitamins, minerals, and chemical compounds. Its crisp texture, mild taste, and nutritious benefits are prized. Salads, sandwiches, and garnishes frequently contain lettuce, a cool-season crop that may be cultivated in a range of regions.

Lettuce has become more well-liked as a result of its adaptability, nutritional advantages, and capacity to improve the appearance and flavor of a variety of foods. Lettuce continues to be a common component in many cuisines throughout the world, adding a fresh and crisp element to meals all year long, whether it be as a garnish, the foundation of a salad, or as the primary ingredient.

History and Origin

Europe and southwest Asia are the places where lettuce (*Lactuca sativa*) originated. The ancient Egyptians had already begun to grow lettuce, although it is unclear whether they did it for eating the leaves, making offerings for the gods, or extracting oil from the seeds. Theophrastus and Herodotus both refer to lettuce from the first millennium BCE. The variety of lettuce is wide and has grown over the past 20 years. Some examples are Butterhead, oak leaf, lollo, Cos (romaine and grasse), European Batavia, and American Batavia (Iceberg).

Other countries of the world began to grow lettuce after it was introduced to Egypt. It made its way to ancient Rome and Greece, where it became well-liked as a staple dish. The cooling and reviving characteristics of lettuce were also valued by the Greeks and Romans, who ate it in a variety of ways including cooked and raw in salads.

Lettuce was brought to the Americas by European conquerors and explorers in the 16th and 17th centuries. The 19th century saw the development of lettuce variants in the United States, including the crisphead lettuce, which later gave rise to the well-known iceberg lettuce.

Nutritional Value

Na is most abundant in lettuce. On the list of commercial vegetables listed in order of nutritional value expressed in unit terms, it comes in at number 26. A leaf contains 900 mg of vitamin A, 100 mg of vitamin

C, 178 mg of choline, 50 mg of calcium, and 28 mg of phosphorus. Cooked lettuce loses the majority of its vitamin C. This vegetable has a useful amount of several minerals, including calcium and iron (Romani *et al.,* 2002). But the nutritional qualities of various lettuce varieties vary. Micronutrient content is higher in leafy varieties than headed ones. Dark green varieties have higher vitamin C, iron, and beta-carotene. Crisphead lettuce is less nutrient-dense than butterhead varieties.

Table 1: Per 100 g edible portion, the leaves contain:

Properties	Amount
Fiber (g)	0.2
Energy Value	50
Water (g)	94
Fat (g)	0.2
Ash (g)	0.7
Protein (g)	1.2

Area and Production

Previously, the main lettuce-producing regions were Europe and North America; but, since the late 20th century, lettuce production and consumption have become widespread. Today, China is the world's largest producer of lettuce, followed by the USA. Currently, the globe produces more than 22 million tons of lettuce over an area of around 1 million hectares (Mou *et al.,* 2008).

India produces 4% of the world's commercial lettuce, placing it third in the world. McDonald's, a global food behemoth, is developing the Nilgiri mountain in South India as a major manufacturing center for iceberg lettuce. As one of McDonald's India's three iceberg lettuce suppliers, Ooty Farms and Orchards joined the McDonald's family. Lettuce is in high demand, and it's expanding quickly. The country's youthful population prefers foreign food, including burgers, pizza, sandwiches, tacos, spring rolls, and other dishes where lettuce plays a key role. As a result, the restaurant business is seeing steady growth.

Botany

Due to cleistogamy, lettuce is an entirely self-pollinated crop. Capitulum (a thick corymbose, flat-topped panicle) is the most common form of lettuce inflorescence. A succession of branches with several capitula forming a panicle are produced by the elongating seed stalk, together with a single terminal capitulum (composite flower), and other flowers. Numerous (10 or more) florets make up each capitulum. The sexual elements of each floret are enclosed in a ligulate (strap-shaped), yellow corolla. The stamens may be seen as the blooms open in the morning. Typically, stamens merge together to produce tubes. Pollen is released via the anthers inside surface. Pollen is clingy and difficult to blow by the wind. There are two stigmatic lobes in style. Self-fertilization is required by the structure of flowers. The fruit is an achene, according to botany.

Taxonomy

Lettuce is classified into the following categories:

Rank	**Scientific Taxonomical Details**
Kingdom	Plantae
Clade	Tracheophytes
Clade	Angiosperms
Clade	Eudicots
Clade	Asterids
Order	Asterales
Family	Asteraceae
Genus	*Lactuca*
Species	*L. sativa*

Land Preparation

Field should be twice plowed and harrowed. Get ready raised beds that are 0.75 m apart and 1 m broad. Include 300 g of vermicompost or carbonized rice hulls per square meter and 1 kilogram of thoroughly decomposed chicken manure. Put mulch or mulching film on rice straws. Use heated, 7–10 centimeter-diameter tin cans to make holes. For crisphead and loose-leaf kinds, the holes may be positioned 30–40 cm apart from the hills, with 2 rows per bed.

Soil Preparation

Lettuce is a vegetable with shallow roots. Because they cover such a tiny area and are so thin and sensitive, their roots are especially vulnerable to drying out. In order to provide for optimum water retention, the soil must be thoroughly tilled and sufficiently drained. With the exception of acid soils, lettuce reacts well to organic manure. In order to meet these needs, you should typically add 1 handful of well-rotted manure to the soil and 5g (1 tsp) of hydrated agricultural lime to each planting hole.

Climatic Requirements

Lettuce is essentially a cool-weather crop, germinating between 15 and 20 degrees Celsius. At temperatures higher than 25°C, germination may be hampered. Additionally, between 15 and 20°C is the perfect temperature range for development. Growing lettuce in the summer is more of a challenge since disease pressure is stronger than in the winter and certain cultivars may bolt because to the hot weather. Its seed germination is affected by both lower and higher temperatures. The germination of seeds is hampered when the soil temperature exceeds 30°C. The leaves are affected and the flavor becomes harsh as the temperature rises above 22°C.

Lettuce Variety

Lettuce is classified into various types, namely

1. **Crisp head or iceberg type lettuce (*Lactuca sativa var. capitata*)**- It is mostly grown in USA. It is heading type with wrinkled, non-wrapper and brittle textured leaves. It is very sensitive to heat and have very good shipping and handling ability.

E.g.

- **Punjab Lettuce No. 1**- It is a non- bunching type of lettuce and bears loses leaves.
- **Great Lakes**- It has large head, bright green leaves with wavy margin and very crispy. It is resistant to tip burn and lettuce mosaic virus. It is usually recommended for cultivation in India.
- **Imperial- 859**- It has medium sized head, light green leaves with soft texture. It is resistant to tip burn.
- **Empire**- It is grown round the year in Europe. It is very crispy and have conical shaped head. It is resistant to bolting.

2. **Leafy or bunching type lettuce (*Lactuca sativa var. crispa*)**- This type of lettuce produces a rosette of leaves. They are early, easy to grow and popular for home gardens. They are not adapted for long-distance transport.

E.g.

- **Slow Bolt**- Its leaves are broad and slightly yellowish to greenish in color. It is resistant to tip burn and usually recommended for cultivation in India.
- **Chinese Yellow**- It is early type of lettuce. Its leaves are tender, light green and crisp.
- **Solan Kriti**- It is a open pollinated selection. It has long leaves, soft and dark green coloured. It is rich in beta carotene and yields around 25t/ha

3. **Cos or Romanian type lettuce (*Lactuca sativa var. longifolia*)-** It is commercially grown in Mediterranean region. Its elongated leaves form loaf shape head.

E.g.

- **Dark Green-** It has narrow leaves with elongated upright head.
- **Florida 83-** It is resistant to lettuce mosaic virus.
- **Paris Island-** It is very popular cos type of lettuce.

4. **Stem type lettuce (*Lactuca sativa var. aspergina*)-** It produces thick stem which are eaten raw after peeling. It is usually cooked for use in stewed and creamed dishes.

E.g.

- **Celtuce-** Its thick stem is eaten after peeling.

5. **Butter head type (*Lactuca sativa var. capitata*)-** It is usually cultivated in greenhouse. It has small, loose heads with oily soft textured leaves. It has better table quality and have more delicious flavour than crisp type.

E.g.

- **White Boston-** It has soft solid heads and inner leaves are oily, soft and have greasy texture.
- **Other examples-** Dark Green Boston, Tom Thumb, Butter Crunch, Red Butter Head, Mary King and Reskia Wonder.

The prominent head types of lettuce grown in India include Iceberg and Great Lakes among the cultivars. Light green, round, and densely covered with lettuce leaves, the head of iceberg lettuce resembles a head of cabbage. Several vitamins and minerals are abundant in lettuce. It is highly helpful for weight reduction attempts as iceberg lettuce has a high-water content (95 percent) and low-calorie content (13 kcal/100g). When consumed during pregnancy, iceberg lettuce helps prevent neural tube problems in newborns as well as heart disease and stroke.

Figure 1.2: Red Leaf Lettuce

Figure 1.3: Heads of lettuce lettuce

Figure 1.4: Iceberg Lettuce

Seedling Production

Each acre needs about 150–200 g of seeds. Set up five seedbeds, each 1 x 10 meters long. Pulverize the soil and add 300 g of carbonized rice hulls per square meter along with 1 kilogram of totally decomposed chicken manure. Make shallow lines 7–10 cm apart after moistening the seedbeds. Thinly plant the seeds after soaking them for several hours in warm water. Mulch with rice hulls, rice straw that has been cut, or coir fiber and lightly cover with soil. For better recovery of pierced seedlings, prick nursery trays with vermicompost during the two-leaf stage. Water often. Offer some partial shade in the dry season and a rain shelter in the wet season. Harden seedlings one week before transplanting by decreasing the frequency of watering and by exposing fully to sunlight to minimize transplant shock. Transplant at three weeks after pricking.

Planting

You may plant lettuce in a variety of containers or on beds. In a home garden setting, lettuce may also be interplanted with herbs and other crops. With a spacing of 30-45 cm x 20-25 cm, the seed rate is 400-500 g/ha. The seeds germinate in around 4-5 days and are suitable for transplanting in 2–3 weeks. When buying seedlings, choose healthy ones that aren't too old.

Transplanting

Remove dirt from the holes that have been created, and then replace it with vermicompost organic fertilizer. Fill all of the holes with water. One seedling should be transplanted each hill; for quicker seedling recovery, push the earth up around the plant. Transplant in the late afternoon on bright days to lessen transplant shock. Hill replacement as soon as possible. Transplanting should take place in the late afternoon. By doing this, the plants will avoid withering brought on by the stress of transplanting. When transplanting:

Weed Management

It is crucial to manage weeds, thus using the approved pesticides, the field should be cleansed of weeds before planting. Weeds increase rivalry and reduce production. field evaluations of pendimethalin and S-metolachlor for weed control in transplanted lettuce are reported. Pendimethalin was evaluated PRE at rates between 0.6 and 6.7 kg ai ha^{-1}, and POST at 1.1 and 2.2 kg ha−1. Both pendimethalin PRE and POST applications were safe to transplanted lettuce and provided similar weed control (Lati *et al.,* 2017).

Fertilizer Application

- Immediately after transplanting, apply 5g (1 tsp) of a granular NPK fertilizer such as 12:24:12 to each plant to encourage root growth. Be sure to place this fertilizer at least 5cm (2 in) away from the plant.
- Rouchaud *et al.* 1984 found that at a normal dose of N, P, K, and S fertilizers the Beta-carotene level in lettuce was 1,666 μg/100 g fresh wt, whereas at a dose of N 40 times lower and lack of other fertilizers it decreased to 1,421 μg/100 g.
- Apply a granular NPK fertilizer, such as 20: 10: 10, at a rate of around 1-3 tsp (5-15g) per plant, depending on the health of the plant, on the seventh, fourteenth-, and twenty-first-days following transplanting.
- A foliar fertilizer, such as 20:20:20, can also be applied on a weekly basis at a ratio of 5g (1 tsp) to 4 liters (1 gal) of water. Apply 125 ml (1/2 cup) of this mixture to each plant's leaves and root region.
- To boost plant vigor and resilience to pests and fungus diseases, apply manure tea and fermented plant juice once a week.

To make dung tea, soak a third of a dry bag of cow or horse manure (30 kg) in a plastic barrel filled with 180 L of water for seven days. Use the tea as foliar fertilizer or a soil drench. Activist plant components should be diced and combined with molasses or brown sugar in equal proportions to make fermented plant juice. After the juice has fermented for a week, remove it and use 1 tbsp/3.785 L of water as a foliar fertilizer.

Irrigation Management

Proper irrigation management is essential for the successful growth and development of lettuce plants. Lettuce has shallow roots, so regular and adequate moisture supply is crucial. The goal is to maintain consistent soil moisture without waterlogging or allowing the soil to dry out excessively. Irrigation should be done evenly, ensuring that water penetrates the root zone without causing runoff. Drip irrigation or soaker hoses are recommended to deliver water directly to the base of the plants, minimizing water waste and reducing the risk of foliar diseases. The frequency and duration of irrigation depend on various factors such as soil type, weather conditions, and stage of plant growth. It is important to monitor the soil moisture regularly and adjust the irrigation schedule accordingly. Mulching can also be beneficial to conserve soil moisture and regulate temperature. By implementing effective irrigation management practices, lettuce growers can optimize water usage, promote healthy plant growth, and achieve high-quality yields.

Harvesting, Yield and Postharvest handling

Crisphead lettuce should be harvested 45 to 60 days after transplantation or when the heads are firm. Harvest loose-leaf lettuce as needed, ideally before it starts to bolt. When there is rain or dew, it is best to avoid harvesting. While leaf type yields more, head type yields are between 100 and 140 q/ha. One carton equals 24 lettuce heads for calculating yield. Sort the heads in the field before immediately packing them in perforated cardboard boxes for large-scale planting. When possible, vacuum cooling to 1°C is recommended. At 4 °C, transport. The storage life of lettuce was shown to be significantly shortened if the surface was left moist, according to Herner and Krahn (1973).

Pest and Diseases Management

The two main pests that affect lettuce are semi-loopers and aphids. Mix 100 g of macerated hot pepper in 16 L of water to make hot pepper solution. Use as a natural pest repellent by adding 1 tbsp of soap. Mulching and sun sterilizing prepared beds for a week before to transplanting can reduce the risk of bacterial rot.

- **Anthracnose (*Microdochium panattonianum*)**

Symptoms- A fungus is typically to blame for the symptoms. Small, water-soaked tan patches that may grow and turn straw-colored are found on the outer leaves. Mature lesions' cores fall off, giving the plant a shot-hole look.

Management- Crops should be rotated, crop trash should be worked into the soil, wild lettuce populations should be controlled near a plantation, and overhead watering should be avoided.

- **Leaf drop (*Sclerotinia minor*)**

Symptoms- Wilting of the outer leaves that progresses to afflicting the entire plant; soft, watery lesions on the leaves; leaves that collapse and lay on the soil surface; and black fungal formations on diseased leaf tissue and the soil surface

Management- Application of fungicides soon following plant thinning, thorough soil cultivation, and crop rotation with non-host species all considerably minimize disease.

Figure 1.5: Lettuce head infected with leaf drop disease

- **Powdery mildew (*Erysiphe cichoracearum*)**

Symptoms- It include older leaves developing a white, powdery fungal growth on top and underneath them, turning yellow or brown, and sometimes even tiny, black fruiting bodies.

Management- As long as temperatures are high enough, disease can be treated by applying sulphur at the first indication of symptoms.

Figure 1.6: Powdery mildew on lettuce

- **Big vein Mirafiori lettuce big-vein virus (MiLBVV)**

Symptoms- It is caused by virus. Veins enlarged and clear; puckered or ruffled leaves; upright outer leaves

Management- Planting resistant varieties is the best method to control MiLBVV.

Figure 1.7: Close-up of lettuce leaf infected with MiLBVV

Future Prospects

According to Baslam *et al.* (2011), consumers are becoming more and more interested in meals that improve wellbeing, lower illness risk, and lengthen life span in addition to meeting their nutritional demands. Customers choose lettuce because it has few calories, is packed with nutrients, and is simple to add to a variety of cuisines. Urban farming techniques like hydroponics and vertical farming are becoming more and more common due to the growth of urbanization and the scarcity of arable land. For these techniques, lettuce works well since it can grow vertically and thrives in nutrient-rich water systems, enabling year-round farming in urban settings. The best growth conditions for lettuce may be found all year long in controlled environment agricultural settings like greenhouses or indoor vertical farms. These innovations enable fine-grained control over variables like temperature, humidity, and lighting, leading to higher yields and more reliable quality, which may help lettuce gain even more popularity. It's critical to remember that these predictions are theoretical and that the future is unknown. But given existing patterns and customer preferences, lettuce is probably going to continue to be a useful and popular vegetable in the years to come.

References

Annalisa Romani, Patrizia Pinelli, Carlotta Galardi, Graziano Sani, Antonio Cimato, Daniela Heimler (2002). Polyphenols in greenhouse and open-air-grown lettuce. *Food Chemistry*. 79 (3) Pages 337-342.

Chadha, K.L. (2019). Hand Book of Horticulture. Second Edition, Volume 1, Indian Council of Agricultural Research, New Delhi. pp. 560-562.

Herner, R.H. and Krahn. T.R. 1973. Chopped lettuce should be kept dry and cold. In "*Yearbook Prod. Mark. Assoc.*" p. 130.

Lati, R., Mou, B., Rachuy, J., Smith, R., Dara, S., Daugovish, O., & Fennimore, S. (2015). Weed Management in Transplanted Lettuce with Pendimethalin and S-Metolachlor. Weed Technology, 29(4), 827-834.

Marouane Baslam, Fermín Morales, Idoia Garmendia, Nieves Goicoechea (2013). Nutritional quality of outer and inner leaves of green and red pigmented lettuces (*Lactuca sativa* L.) consumed as salads. *Scientia Horticulturae*. 151(103-111).

Mou B, Prohens J, Nuez F, Eds (2008). Handbook of plant breeding, vegetables I, Asteraceae, Brassicaceae, Chenopodicaceae, and Cucurbitaceae. *New York: Springer* pp. 75-116.

Rouchaud J, Moons C, Meyer JA, Benoit F, Ceustermans N, van Linden F (1984). Effects of soil fertilization and covering of the culture with plastic film on the provitamin A content of early lettuce. *Plant Soil*; 80: 139-41.

20

Advances in Production Technology of Asparagus

Jagraj Singh[1*], Satya Prakash[2], Amit Kumar[3], Mohit[4] and Chetan Chauhan[5]

[1,2&4] Department of Vegetable Science

[3]Department of Fruit Science

*[5]Department of Floriculture and Landscape, College of Horticulture, Sardar Vallabhbhai Patel University of Agriculture & Technology, Meerut-250110 Uttar Pradesh *Email: bhullarjagraj1313@gmail.com*

Introduction

The asparagus (*Asparagus officinalis* L.) is an herbaceous perennial plant and known as Shatawar in Hindi. It is grown for tender succulent shoots known as '*Spears*' which are used as vegetable and in the preparation of soup. Once the plant established, it continues to produce yield for 10-15 years. The juice obtained from spears contain a white crystalline substance called '*Asparagine*' which has diuretic properties. Asparagus is a type of vegetable that is commonly enjoyed for its unique taste and nutritional benefits. Asparagus has a distinctive spear-like shape, with a green, purple, or white colour depending on the variety. Asparagus is a low-calorie (20 Kcal) vegetable that is rich in protein (2.2 g), vitamins, including vitamins B1 (0.143 g), Vitamin B2 (0.141 g), Vitamin C (5.6 g), Vitamin E (1.13 g) and Vitamin K (41.6 µg), as well as minerals including calcium (24 mg), iron (2.14 mg), potassium (202 mg), zinc (0.54 mg), sodium (2.0 mg) and magnesium (14 mg) per 100 g raw asparagus. It is also a good source of antioxidants, which help to protect the body against damage from free radicals. Asparagus can be cooked in a variety of ways, including boiling, steaming, roasting, or grilling. It is a popular ingredient in many dishes, such as salads, soups, pastas, and stir fries. Asparagus is also often served as a side dish, either on its own or with other vegetables.

Botany

Asparagus is a perennial, dioecious herb in the Liliaceae family, which is native to Europe, North Africa, and Western Asia. It is widely cultivated as a vegetable crop in many parts of the world. The asparagus plant has a fern-like appearance, with long, thin, and feathery leaves that grow from a central crown. The plant can grow up to 1.5 meters tall and has a fleshy, underground root system called a 'crown', which can live for up to 15 years. Asparagus is dioecious, which means that individual plants are either male or female. The male plants produce small, inconspicuous flowers that are yellowish-green in colour, while the female plants produce larger, showier flowers that are pinkish-white. The underground portion is made up of rhizomes or crowns and their lateral modifications, as well as fleshy and fibrous roots. The crowns carry the buds, which have been organised in a compact sequence on the upper regions. The aerial stems that form the delicious spears emerge from the rhizomes the previous year.

Taxonomy

The asparagus (*Asparagus officinalis* L., 2n=2x=20) plant belongs to the Liliaceae family and is classified under the genus Asparagus. Within the Asparagus genus, there are several species of asparagus, including *Asparagus officinalis*, *Asparagus racemosus*, *Asparagus sprengeri*, and *Asparagus setaceus*. The most common species of asparagus is *Asparagus officinalis*, which is widely cultivated as a vegetable crop. Within this species, there are several varieties, including green, purple, and white asparagus.

Soil and Climate

Being a perennial crop, root system of asparagus may develop 10 feet's deep, so soil should be free from any hardpan. Generally, it requires deep, fertile and well drained sandy loam soil with rich in high organic matter. The optimum soil pH should be in between 6.5-7.5. Asparagus cannot tolerate to extreme acidic conditions.

Asparagus is a temperate season vegetable and bears excellent crop during a temperate range from 16-24°C for most of the growing season to produce quality spears. It can tolerate frost up to some extent but continuous freezing is harmful for its growth. Spear initiation occurs at the soil temperature more then 10°C. Root and shoot development are reduced at temperature below 13°C and above 29°C while, Spear elongation is faster at air temperature.

Field Preparation

On deep ploughing followed by 2 -3 cross harrowing's are necessary to make the soil friable and loose. Planking should be done just after each harrowing to level the field.

Commercial Varieties

The common varieties are Perfection, Mary Washington, Martha Washington, Colossal, Donald's Elmira, Palmetto, Colossal, UC 72, Brock Imperial 84, Reading Giant, Jersey Queen, Jersey Giant, New Jersey and Improved Perfection are grown in different countries. Varieties such as UC 66, UC 72, Selection-831 are also grown in Jammu & Kashmir in India.

Perfection: It is recommended by IARI, New Delhi. It is an early, uniform, productive variety, delicious with high food value. The spears of this variety are early, large, green, uniform in maturity, succulent and light tipped. Average yield of this variety is 80-100 q/ha.

Selection-841: Plants of this variety are uniform, medium, bush type, and productive. The spears are 15-20 cm long, succulent, tender, green with better flavour and suitable for soup preparation. Yield potential of this variety is 90-110 q/ha.

Palmetto: It has a somewhat different appearance than other types, is significantly resistant to rust, and is widely planted throughout America.

Jersey Queen: It is a cultivar cloned from Mary Washington. It induces high-quality spear with a high yield.

Gijnilm: This is a hybrid asparagus cultivar. It is still in the early stages of manufacture, with medium, thick, straight spears of outstanding quality.

Seed rate

There are two methods for growing asparagus:

1. **Raising by Seed:** About 3-4 kg of seeds are required to raise seedlings for planting one hectare area. Before planting, seeds are soaked in water for 2-3 days. Seeds are sown in nursery beds at a spacing of 20-25 x 7-10 cm followed by 1.5- 2.0 cm depth. Seedlings become ready for transplanting in main field 8- 10 months after sowing.
2. **Propagated by Crowns:** Select one year old well grown crowns of big size containing good growth and having large buds. Crowns are separated from mother plant before buds sprout and planted within a few hours after separation.

Sowing and Transplanting

The optimum sowing time of asparagus in hills from March to May and in plains from July to November. Transplanting of seedlings in the bottom of furrows made in the main field at a spacing of 75- 90 x 45 cm. Transplant the crowns in 15- 20 cm deep furrows spaced at about 1.5 – 2 meter keeping the distance of 45-60 cm. covers the crown properly with soil to 2-3 cm above ground level and press around the crowns firmly.

Nutrition Management

At the time of field preparation, well decomposed FYM @ 300-400 q/ha should be mixed in the field. In addition to this 100-120 kg nitrogen, 80-100 kg phosphorus and 60-80 kg potassium is also applied in the field. The half dose of nitrogen and full dose of phosphorus and potassium just before the planting of seedlings or crowns. Rest half amount of nitrogen may be applied in 4- 5 split doses during harvest season just after cuttings. The requirement of nitrogen is more during first year and after that decreased to about 50 percent.

Irrigation Management

Irrigation is essential to release drought stress, especially during first two years after transplanting of crowns. First irrigation should be given at the time of transplanting of crowns. After that irrigation should be given at 10-15 days interval depending upon the crop requirement or soil moisture. Excess water should be avoided because asparagus can not tolerant the water logging condition for long time, so proper drainage should be provided in the field.

Weed Management

Field should be free from weeds by timely shallow hoeing between the rows without damaging the spears. Weeds can be effectively control by the application of herbicide like metribuzin @ 1 kg/ha at pre-emergence and early post emergence stage for successfully controlling the broad-leaved weeds.

Blanching

Blanching is practised to blanch the young spears and get white asparagus for canning purpose. Because blanched (whitish) spears are preferred for canning. In blanching cover the plants with soil up to a height of 25- 30 cm before harvesting the green spears for fresh market.

Mulching

It has been revealed that mulching the asparagus bed helps to reduce weeds, manage soil moisture and temperature, and promotes better development and higher early output with a higher overall crop yield. Mulching with polythene or polyvinyl carbonate film increased soil temperature and asparagus production by 25-30%. The use of 0.05 mm thick transparent polythene film elevated soil temperature by 2-30 degrees Celsius and yielded 17% greater early yield than a non-mulch crop.

Physiological Disorders of Asparagus and their Managements

Asparagus can be affected by various physiological disorders, including:

Purple Spear- It is a physiological disorder that causes the spears of asparagus plants to turn purple or reddish-brown. It is caused by exposure to sunlight or low temperatures and does not affect the quality of the crop.

Hollow Heart- Hollow heart is a physiological disorder that causes the interior of asparagus spears to become hollow. It is caused by uneven growth or temperature fluctuations during spear development and does not affect the quality of the crop.

Major Insect- Pest and their Management

Asparagus can be susceptible to various insect pests. Some of the most common insect pests of asparagus include:

Asparagus Beetle- The asparagus beetle (*Criocersis asparagi*) and the twelve-spotted asparagus beetle (*Criocersis duodecimpunctata*) attack asparagus. Plants get eaten by both adults and larvae. Adult beetles feed on spear tips in the spring, reducing crop quality. During the summer, they infect the ferns and disrupt the crown's food supply. The symptoms caused by adults are identical to those caused by common beetles, however the larva stage causes less crop harm because it feeds on fruits or berries.

Control- Spraying with carbaryl 50 WP @ 3g/litre of water at least twice at 7 days interval can control the beetles.

Asparagus Aphid (*Bracycolus asparagi*)- Aphids are little, soft-bodied insects that feed on asparagus sap. They are usually found on the undersides of leaves and can be green, yellow, black, or brown in colour. Asparagus aphid feeding results in busy, stunted new growth known as 'witched broom'.

Control- Spray the crop with Imidacloprid 17.80 SL @ 0.5-1 ml/litre of water.

Garden centipede (*Scutigerella immaculata*)- The insect bores small, round holes in the spears beneath the ground surface, rendering them unfit for ingestion.

Control- Flooding of water in scattered infested field is advisable for its control. Soil application of carbofuran @ 15 kg/ha is also effective to control the pest.

Major Diseases and their Management

Asparagus is a hardy and long-lived perennial plant that is grown for its edible young shoots or spears. Like all plants, asparagus can be susceptible to a range of diseases. Here are some of the common diseases that affect asparagus plants:

Crown Rot- This is the most devastating disease affecting asparagus crop worldwide. It is also known as asparagus wilt and root rot disease. *Fusarium oxysporum* sp. *asparagi* and *Fusarium moniliforme* are soil-borne fungi that cause it. Poor soil drainage or overwatering are common causes of the disease. It causes delayed crown degradation and a reduction in spear size and quantity, lowering the yield. The vascular tissue of shoots and storage roots in infected crowns has a reddish-brown colour.

Control- It can be controlled by use of vigorous growing varieties, use of disease-free seeds or crowns, proper nitrogen application, timely irrigation, high phosphorus and potash level, weed and insect control help to reduce plant stress thus reduce the infection of *Fusarium*. Apply fungicides like chlorothalonil or azoxystrobin during the early stages of spear growth.

Asparagus Rust- It is caused by fungus *Puccinia asparagi*. The disease grows in hot, humid areas. This fungal disease develops rust-coloured pustules on the plant's stems and fronds. It can cause early defoliation and yield loss. The spore can be transmitted to connected plants by wind and rain.

Control- Even before infection occurs, it can be managed by dusting fine sulphur at a rate of 24 kg/ha. Spray benlate at 0.1% at the start of plant growth. Rust-resistant cultivars have been recommended. Good irrigation management, wide row spacing, and the use of fungicides may help to decrease asparagus rust infestation. Use fungicides like tebuconazole or propiconazole during fern growth to prevent spore production.

Spear Rot- It is caused by soil born fungus *Phytophthora megasperma* var. *sojae*. This fungus attacks asparagus particularly in the soil that are saturated due to poor drainage, heavy rainfall or excessive irrigation.

Control- To prevent spear rot disease, it is important to maintain good soil drainage, avoid overwatering, and keep the asparagus bed free of weeds and debris. Additionally, planting disease-resistant varieties and rotating crops can also help prevent disease. Use fungicides like tebuconazole or propiconazole during fern growth to prevent spore production.

Asparagus Purple Spot- It is caused by *Stemphylium vesicarium* and can arise in cool and damp weather. It generates oval-shaped purple spots on the spears, lowering crop value. Spots extend and create greyish white, slightly depressed centres inside the lesions as they progress.

Control- Good sanitation, removal of asparagus debris of previous season from the field is helpful in controlling the disease. Apply fungicides like thiophanate-methyl or cyprodinil during the spear development phase.

Yield

The yield is obtained from the third year onwards and continues to produces economic yield for about 10-15 years. The yield increases up to 6-7 years, remain constant up to 12 years and then starts to gradually decline. However, the crop between 4-10 years produces best quality spears. The average yield of asparagus is about 25- 40 q/ha.

Harvesting and Post-harvest Management

The first cutting should be done two years after planting. Spears (10-15 cm size) are cut with the help of asparagus knife just above the soil surface (3-5 cm) at an angle of 45°C are should be taken at the time of harvesting so that young sprout should not be damaged. The spears should not be harvested one year after planting of crowns because during first year the growth of rhizomes occurs. The harvesting should be done during morning hours at alternate days and quickly sent to market.

Post-harvest management is a crucial aspect of asparagus production that determines the quality, shelf life, and market value of the produce. Proper handling and storage of asparagus spears are essential to prevent moisture loss, decay, and deterioration of quality. Here are some guidelines for post-harvest management of asparagus spears-

Harvesting: Asparagus spears should be harvested when they are young and tender, usually 10- 15 cm long, and have tightly closed tips. Spears should be cut close to the ground using a sharp knife or blade to avoid damage to the crowns.

Sorting and grading: After harvesting, sort the spears according to size, colour, and quality. Grade them according to the USDA standards, which include four grades: Extra, Standard, Large, and Jumbo.

Cooling: Asparagus spears should be cooled as soon as possible after harvest to prevent quality deterioration. The ideal storage temperature is 32°F to 36°F (0°C to 2°C) and a relative humidity of 90% to 95%.

Packaging: Asparagus spears should be packed in plastic bags or containers to maintain their moisture content and prevent mechanical damage during transportation.

Storage: Asparagus spears should be stored in a cool and humid environment. They can be efficaciously stored at 0- 2°C with 95% RH for 2-3 weeks in good conditions. Speared stored in wet tissue paper may keep fresh and firm even after 13-16 days of storage. At room temperature, the spears become tougher if stored even for short period.

Pre-cooling: Pre-cooling the asparagus spears in a hydrocooler or forced-air cooler can improve their shelf life and quality.

Handling: Asparagus spears should be handled gently to prevent damage to the tips and sides. Avoid dropping or stacking the spears too high.

By following these guidelines, asparagus growers can ensure the quality and freshness of their produce, leading to increased market value and profitability.

Future Prospects

Asparagus cultivation has a promising future due to its increasing demand in the global market. Asparagus is a highly nutritious and low-calorie vegetable that is rich in vitamins, minerals, and dietary fibre, making it a popular choice among health-conscious consumers. In addition, asparagus is versatile and can be used in a wide range of culinary applications.

Some of the future prospects of asparagus cultivation are:

1. **Growing demand**: The demand for asparagus is expected to continue growing due to its increasing popularity and its reputation as a healthy vegetable.
2. **Health benefits:** Asparagus is a rich source of vitamins, minerals, and dietary fibre, which are essential for maintaining good health. As more people become health-conscious, the demand for asparagus is likely to increase.
3. **Diverse uses:** Asparagus is a versatile vegetable that can be used in a wide range of culinary applications. It can be grilled, roasted, sautéed, or used in soups and stews. As consumers look for new and creative ways to use asparagus in their cooking, the demand for this vegetable is likely to increase.
4. **Sustainability:** Asparagus cultivation is a sustainable agricultural practice that can help to protect the environment. As more consumers become environmentally conscious, the demand for sustainably grown asparagus is likely to increase.
5. **Advancements in technology**: Advancements in technology are helping to improve the productivity and efficiency of asparagus cultivation. New cultivation techniques, such as hydroponics, can help to increase yields and reduce water usage, making asparagus cultivation more profitable and sustainable.

Overall, the future prospects of asparagus cultivation are bright, and the demand for this nutritious vegetable is expected to continue to grow in the coming years.

References

Chadha, K.L. (2019). Hand Book of Horticulture. Second Edition, Volume 1, Indian Council of Agricultural Research, New Delhi.

Fageria, M. S., Chaudhary, B. R. and Dhaka, R. S. (2009). Production Technology of Vegetable Crops. Kalyani Publishers, New Delhi. pp, 301-304.

http://ecoursesonline.iasri.res.in/mod/page/view.php?id=12140

https://agritech.tnau.ac.in/horticulture/pdf/tech_bulletin/international/Asparagus%20 Production.pdf

https://vikaspedia.in/agriculture/crop-production/package-of-practices/medicinal-and-aromatic-plants/asparagus-racemosus

https://www.agrifarming.in/asparagus-cultivation

https://www.gov.nl.ca/ffa/files/FFA-Crop-Guide-Asparagus.pdf

https://www.researchgate.net/publication/327282704_Scientific_Cultivation_of_Asparagus

Kaur, R., Jaidka, M. and sepat, S. (2018). Scientific Cultivation of Asparagus, Advances Vegetable Agronomy, *Director IARI, New Delhi & ICAR- DARE, Ministry of Agriculture, Government of India New Delhi–110 001*, India, pp. 271-279.

Singh, D. K. (2007). Modern Vegetable Varieties and Production Technology, *International Book Distributing Co.* pp. 359-362.

Thamburaj, S. and Singh, N. (2022). Textbook of Vegetables, Tuber crops and spices. ICAR. New Delhi, pp. 426-428.

21

Advances in Production Technology of Globe Artichoke

Bibarie Mech[1], Chhail Bihari[2], Elizer R Marak[3], Sajeel Ahamad[4] and Abdul Rahman M[5]

[1] *M.Sc. Research Scholar, Department of Horticulture, Nagaland University, School of Agricultural Sciences, Medziphema, Nagaland*

[2,3&5] *Ph.D. Research Scholar, Department of Horticulture, Nagaland University, School of Agricultural Sciences, Medziphema, Nagaland*

[4] *Ph.D. Research Scholar, Division of Food Science & Postharvest Technology, Indian Agricultural Research Institute, New Delhi, India*

Email: cbrathour500@gmail.com

Introduction

The Globe artichoke is a herbaceous perennial crop that is grown for its tender, edible, immature flower buds. It is botanically known as (*Cynara scolymus* L.) and having chromosome number 2n=2x=34. The globe artichoke belongs to the family Compositae, and it originates in the Mediterranean Basin, where it was most likely introduced as a crop in about the first century AD (Sonnante *et al.*, 2007). The globe artichoke is presently grown as an annual crop or a perennial crop in different regions of the world. The increasing demand for globe artichokes as a source of pharmaceuticals and modern functional foods has created new challenges for producers, necessitating the concurrent development of new production hubs and the use of cutting-edge technology in established growing regions that can support the global market.

Origin and Distribution

The globe artichoke, scientifically known as *Cynara scolymus* L., is a herbaceous perennial cultivated primarily for its edible flower buds. These buds are harvested when they are still young and tender. The artichoke originates from the Mediterranean region, where it was probably originally cultivated during the first century AD (Sonnante *et al.*, 2007). The primary globe artichoke-growing location is still the Mediterranean region. Italy holds a dominant position in globe artichoke production, with the coastal plains of southern regions such as

Sicily, Apulia, Sardinia, and Campania being key cultivation areas. Spain takes the lead as the largest exporter of globe artichokes, while France serves as an importer of both fresh and preserved globe artichoke products (Lombardo *et al.*, 2017; Bianco, 2005). Apart from Southern Europe and North Africa, globe artichokes are cultivated in several other countries, including China, the USA (especially California), Argentina, Peru, Chile, and Brazil. In countries like Argentina, Peru, and Iran where the production of the globe artichoke has not had a long heritage, different timing and production techniques might result in large yields (Macua, 2007). In India, artichoke cultivation is practised on a small scale. It is grown in Himachal Pradesh, India, as a luxury vegetable. In European nations, it is an important vegetable crop, nevertheless. Around 1,300,000 t of buds from more than 120,000 hectares of globe artichokes are grown on more than 25 different continents. This area is 2.5% of the total land used to grow vegetables worldwide. With more than 35% of global production, Italy holds the top artichoke-producing country globally. With 19% of global production, Spain comes in second.

Botany and Taxonomy

The globe artichoke (*Cynara scolymus* L.) is a perennial herbaceous plant with edible flower buds that is susceptible to the cold. The 4 to 5 feet tall, 5 to 6 feet wide plants have a silvery green colour. The lateral stems and the terminal portion of the main stem are wherein the flower buds develop. Each unopened flower bud has a spherical, somewhat elongated shape, is 3 to 4 inches in diameter, and is deep green in colour. A purple-blue blossom is encircled by many pointed, leathery, green bracts. Along with the fleshy centre of the artichoke, on which the flower and bracts are borne, the base of each bract is the fleshy edible part. A receptacle containing wide, thick, green, violet bracts (the edible part), around a centre cluster of violet-blue florets, makes up a flower head. The plant's leftover buds open to produce 6-inch purple-blue flowers. These are used in flower arrangements after being dried. In addition to C. cardunculus, the genus Cynara has seven other species and is indigenous to the Mediterranean Basin. It is a member of the botanical family Asteraceae (Compositae). The other two kinds of *C. cardunculus* are believed to have their origins in *C. cardunculus* var. sylvestris. The allogamous species of artichoke is primarily vegetatively reproduced by "stumps" (basal stem fragments with attached root sections) or "ovoli" (axillary buds separated from the stumps), though some cultivars can also be reproduced by seeds, suckers, offshoots, stumps, or dried shoots from the parent plant.

Nutritional Value

Globe artichokes are an excellent source of potassium, which helps the body maintain fluid balance, and folate, which helps the body produce healthy red blood cells. Additionally, the artichokes provide moderate levels of phosphorus for healthy bones and teeth, vitamin K to aid in quicker wound healing, and other minerals like manganese and magnesium.

Nutritional composition (per 100 g of edible portion)

Contents	Amount	Contents	Amount
Moisture (%)	84.94	Riboflavin (mg)	0.066
Carbohydrates (g)	10.51	Niacin (mg)	1.046
Proteins (g)	3.27	Ca (mg)	44
Fat (g)	0.15	P (mg)	90
Fibre (g)	5.4	K (mg)	370
Energy (kcal)	47	Na (mg)	94
Vitamin-C (mg)	11.7	Mg (mg)	60

Importance and Uses

The globe artichoke, scientifically known as Cynara cardunculus var. scolymus L. Fiori, holds a significant place in the Mediterranean diet as a staple ingredient. Globe artichoke are mostly cultivated in Mediterranean regions for a variety of uses, including traditional human consumption as well as alternative applications. These versatile crops are grown both for their use as agricultural food crops and for non-food purposes. The edible sections of artichokes include high-quality inulin, fibres, and minerals in addition to being one of the diet's best sources of polyphenols with a high bioavailability. Pharmaceutical artichoke leaf extracts also possess choleretic and hypocholesterolemic qualities. The globe artichoke has a long history of use in traditional medicine, spanning thousands of years, primarily because of its well-established therapeutic effects. including anticarcinogenic, hepatoprotective, antioxidant, antibacterial, anticholesterol, urinative, and glycemic control (Kraft 1997; Gebhardt 1997; Fantini *et al.*, 2011 and Coinu *et al.*, 2007). Extensive clinical research has consistently shown that the bioactive characteristics of globe artichoke are attributed to its high concentration of polyphenolic chemicals, including hydroxycinnamates and flavonoids, predominantly found in its flower heads and leaves. One of the highest overall antioxidant properties for vegetables has been found in artichoke flower heads, which are directly connected to their polyphenolic content. Mycorrhizal inoculation, as an agronomic factor, can potentially influence the secondary metabolism of plants, specifically in the synthesis of compounds that possess health-promoting properties. According to new findings, mycorrhizal artichoke plants' leaves and flower heads now have much increased levels of total phenolic content and antioxidant activity. Further research has been conducted to explore potential applications of globe artichokes as a source of fibers for reinforcing composite materials. Positive results were obtained from the investigations conducted on biodiesel production and the utilization of seed oil as a food source. The ability to serve several uses may hold the key to the future global proliferation of these crops.

The edible young flower heads of the globe artichoke, considered a delicacy, can be enjoyed in multiple forms: fresh, frozen, or canned. According to Conti *et al.* (2005), the edible portions of the artichoke are in the bottom parts of the bracts and the receptacle. The artichoke plant is not only valued for its edible parts but also for its leaves, stem, and roots, which are utilized for various purposes as outlined below:

a) It can be used as forage for livestock.
b) Inulin can be extracted from it.
c) It can serve as feedstock for the extraction of secondary metabolites.
d) The plant's lignocellulosic biomass is utilized for energy and paper pulp production.
e) Its seed oil is a valuable resource for biodiesel fuel production.
f) Additionally, it is employed in the production of alcoholic beverages.

Varieties

In the world, there are about 140 varieties of artichokes, of which 40 are grown commercially. Globe artichokes are also common houseplants, frequently sold under varieties such as Green Globe, Chianti, Mercury, Big Heart, King, and Siena.

- The vegetatively propagated perennial cultivar Green Globe accounts for more than half of the artichoke output in California.
- Seeded artichoke varieties include Imperial Star, Desert Globe, Emerald, Big Heart, and Green Globe Improved. A few bigger companies have created their own varieties.

Classic Green Globe

The classic green globe, sometimes known as just the globe, has a heart and bottom with a buttery flavour and ample of flesh at the base of the petals. This artichoke, which has a diameter of 3 to 5 inches and is often planted as a perennial, was initially imported to California from Italy but resembles the French camus de bretagne, a summer choke that grows in Brittany, in terms of both form and flavour.

Chianti

The chianti, a four-inch-wide, green artichoke with a touch of maroon on the leaves and a classic form, is related to the famous Italian romanesco, just as the mercury.

Imperial Star Artichoke

The most well-liked homegrown type is the imperial star artichoke, with 4-5 inches broad buds. Though slightly sweeter, it tastes like the green globe type.

Mercury

The little mercury artichoke, which is typically three and a half inches in diameter and has a reddish-violet hue and a distinctive rounded top, is sweeter than many other artichokes. The Italian romanesco is where the mercury and baby anzio both come from

Big Heart

Developed in the mid-1980s by a California grower named Rusty Jordan, the big heart is aptly named. It is approximately 500 grammes in weight and has a substantial, meaty foundation. This green, 3 ½-5 ½" giant–the first patented annual artichoke grown from seed.

King

The leaf tips of the blocky, brightly coloured king have conspicuous green dots. The king normally measures 4 inches in diameter, weighs more than 500g during peak season, and is bred from romanesco variety mixed with other Italian artichoke strains.

Siena

The oblong siena has a wine red colour with a slight choke. It has a diameter of around 4 inches and was created through a breeding programme in central Italy. This little artichoke, which takes a while to develop and is still only produced in modest amounts, often weighs less than a pound and has a heart soft enough to be eaten raw.

Carciofo Romanesco Artichoke

The leaves of this variety are a deep purple with just a little of green. It has a big heart, nutty flavour, and is delicate and thornless like most artichokes.

Tempo Artichoke: This hybrid variety has 4-5 inches wide buds with purple tinge on bracts.

Soil and Climate

Although the globe artichoke may grow in various types of soil, it thrives in deep, rich, well-drained soil. The most ideal soil is almost sandy. The roots must have good drainage to avoid rotting, especially in places where they will spend the winter. But during the hot summer months, the soil must also be able to hold onto water for a long enough period of time for the roots to absorb it. For vernalization, the artichoke (Cynara cardunculus var. scolymus L.), which is suited to a Mediterranean environment, has to accumulate 200 to 500 hours below 7.2°C. The ideal daytime and nighttime temperatures for artichoke cultivation are 20-24°C (68-75 °F) and 12-14°C (54-57 °F), respectively. The plants can withstand temperatures between -3 and -5°C (26.6 and 23 °F), however, the artichokes cannot survive in temperatures below 1°C (33.8 °F). Bract cells begin to die at temperatures between 0°C (32 °F) and -3°C (26.6 °F), and necrosis of leaves and stems is a regular occurrence once temperatures reach -5°C (23 °F). Cutting back the plant's foliage in the autumn, when it is dormant, and mulching thickly around the plant, burying it up to 30-46 cm (1-1.5 ft) in dirt and straws, is a popular practice in locations where frost threat exists. Although mulching can make plants more tolerant to cold temperatures, no plant can survive temperatures below -9 °C (15°C).

Propagation and Planting

There are different methods to reproduce artichokes, but the three that are most frequently used are seed, offshoots (side shoots), and root-crown division (underground shoots with apical and lateral buds, or "ovoli").

Methods of Propagation

Vegetative Propagation

The globe artichoke is cultivated as a perennial in the Mediterranean basin (Italy, Spain, France, Greece, Turkey, Morocco, and Tunisia. The most often used method of vegetative propagation is the employment of offshoots, which is followed by propagation from stumps (ceppaie) and subterranean shoots with apical and lateral buds, known as ovoli in Italy (Morello *et al.*, 2005). The rootstock used for propagation is manually divided into many sections in the USA and some regions of France. These portions each contain lateral buds from which new plants grow. Planting lateral branches that have been cut off from the mother plants and have fully established roots is a common practise in France, Spain, and Italy. At the same time, it is typical to sow lateral branches with buds that are still dormant across much of Italy.

In-vitro Propagation

A different method to produce abundant, healthy, high-quality, and homogeneous vegetative material is globe artichoke micropropagation. The use of in vitro propagation of globe artichoke as a means of improving its rate of multiplication was reported. Mycorrhizal symbiosis has been employed in globe artichoke micropropagation in recent years to boost plant growth and survival rates by lowering transplanting-related stressors (Campanelli *et al.*, 2013; Ruta *et al.*, 2016). Micro-propagated cultivars are now widely used in European nations because of new effective in vitro protocols, where the high cost of planting material has been offset by improved field performance that satisfies consumer demands (Tavazza *et al.*, 2016; El Boullani *et al.*, 2017).

Seed Propagation

The lack of cultivars suitable for an annual cycle that would provide balanced output and quality of heads was a barrier to the widespread cultivation of annual globe artichoke crops for several years (Virdis *et al.*, 2014). Italian globe artichoke breeding programmes attempt to develop potential seed-propagated cultivars in response to the morphological and financial drawbacks of the vegetative multiplication approach. With hybrids that are not Mendelian F1 and the uniformity resulting from crossings between two pure lines, breeding confronts significant difficulties. According to Pagnotta *et al.* (2016), the globe artichoke exhibits severe inbreeding depression. Out of 500 crosses, about 30 were examined for morphological and agronomic traits and uniformity; some of these were registered, including Romolo. It was determined that hybrid homogeneity was the most crucial factor for morphological and qualitative attributes. Due to its strong processing yield and low total polyphenol content, Opal F1 and Madrigal F1 delivered the finest quality heads for the fresh-cut and processing industries among the novel hybrids. According to (Bonasia *et al.*, 2010), Tempo F1 was considered a potential source of natural antioxidants for the food and pharmaceutical industries. The time from transplanting to maturity can vary from 4 to 6 months, depending on when transplanting takes place. In several regions of the state, seeded artichokes may be cultivated. Annual artichokes are sometimes planted directly in the ground in certain regions of the state, but for the most part, they are transplanted to minimise weed and disease issues. Plains are sown between August and October, whereas hills area sown between March and May.

Grafting

Grafting plays a crucial role in managing *Verticillium* spp. in globe artichoke cultivation. Evaluations have been conducted to assess the resistance of wild and cultivated cardoon accessions against *Verticillium* spp. (Pandozy *et al*., 2015; Ciccarese *et al*., 2012). While grafting is an effective technique, careful consideration of the grafting method and synchronization of sowing times are essential. Among the assessed techniques, splice grafting has shown favorable results. (Temperini *et al*., 2013; Trinchera *et al*., 2013) evaluated using the method of grafting globe artichokes onto cultivated cardoon rootstocks, and found that it increases yield and reduces the incidence of *Verticillium* spp. The splice grafting technique has proven to be highly effective. However, the high costs and labour demands associated with grafting have limited its widespread adoption. Nonetheless, grafting presents a sustainable solution for preserving traditional globe artichoke landraces and cultivating high-quality genotypes in their natural habitats.

Propagation of the crop is achieved using suckers or off-shoots from the old rootstock, as well as by dividing the old crown into pieces along with a portion of the stem. When the suckers or off-shoots reach a height of 30-45cm, they are removed.

The crop can also be propagated using seeds, although the resulting plants are highly variable and are not widely favoured. For seed propagation, the cultivation begins by preparing raised seed beds. Seedlings that are 10-15 cm tall are utilized for planting. The width of the bed usually falls within the range of 72 to 80 inches (1.8 to 2.0 m), allowing for a single row of plants with a spacing of 30 inches (76 cm) between them. The duration required for artichokes to reach maturity after transplantation, ranging from 4 to 6 months, depends on the timing of the transplantation itself. Seeds can be used to cultivate artichokes in different regions. While certain areas practice direct seeding, transplanting is generally preferred to effectively manage weed and disease concerns. In plains, sowing is typically performed between August and October, while in hilly regions, it is carried out from March to May.

Manures and Fertilizers

Perennial artichoke plantations require moderate nitrogen levels. Organic matter, such as manure at 22 t/ha, is applied to enrich soil quality (De Vos, 1992). To reduce nitrogen fertilizer dependency, mineral-organic fertilizers are recommended (Ierna and Mauromicale, 2013). Initially, young transplants need 22-34 kg/ha of nitrogen, providing enough for the first month. As the crop matures, nitrogen requirements increase. For instance, 6 kg/ha/week in early stages and 12 kg/ha/week as plants grow larger. A total seasonal fertigation of 144-180 kg/ha of nitrogen, assuming efficient drip irrigation, is usually sufficient. Late-season planting after other crops benefits from residual nitrogen. The pre-sidedress soil nitrate test (PSNT) measures soil nitrogen availability, with levels above 20 ppm considered adequate. Recommended mineral rates of nitrogen, phosphorus, and potassium in perennial artichoke plantations typically ranges from 168-336 kg/ha, 24-48 kg/ha, and 28-93 kg/ha, respectively. In France, nitrogen recommendation ranges between 150-280 kg/ha (Ryder *et al.,* 1983). Nitrogen, phosphorus, and potassium are applied after harvest, with additional nitrogen applied through top dressing and irrigation.

Cover Crop and Crop Rotation

Conventionally, globe artichoke farming in the Mediterranean region has relied heavily on monoculture practices and the intensive application of nitrogen fertilizers, which has raised concerns regarding sustainability. A study conducted by Lenzi *et al.* in 2015 investigated the utilization of globe artichoke as a cash crop and cover crop in organic systems. The results showed a yield of 7 tons/ha for artichoke heads and 50.3 tons/ha of biomass when used as green manure. Spanu *et al.* (2017) reported a cropping system emphasizing soil fertility management through cover crops, rotations, and abandoning chemical fertilizers. Soil quality recovery was achieved by incorporating fertility-building legumes as catch/cover crops and ploughing crop residues into the soil. Deligios *et al.* (2017) proposed a biannual rotation system with cauliflower and cover crop to optimize nutrient fluxes for conventionally grown artichokes. Implementing sustainable horticultural systems have been developed to adapt to specific local conditions and incorporate crop rotation practices shows promise in reducing synthetic fertilizer use and enhancing globe artichoke productivity in various regions. This approach facilitates a transition toward sustainable practices in traditional globe artichoke cultivation.

Irrigation Management

Artichoke cultivation requires ample water compared to other vegetables due to its extensive leaf area and lengthy production cycle (Leskovar and Xu, 2013). Thus, regular irrigation is crucial for establishing the crop successfully. Research in Mediterranean countries suggests that drip irrigation is the most efficient method for maximizing artichoke yield and water use efficiency. Careful management of irrigation is necessary to avoid soil saturation and prevent root rot. In perennial artichoke fields, winter ditches are dug to drain excess rainwater. Initially, overhead sprinklers are used for irrigation, while subsurface drip systems buried at 30 to 34 cm beneath the soil surface are employed later on. Some growers opt for a single or dual buried drip line one on each side of the plant row. In hilly regions, even-flow irrigation tape ensures uniform water distribution. Annual plantings usually rely on overhead sprinklers. Most farmers prefer irrigating artichokes during early morning hours or the late evening period to minimize transpiration.

Weed Management

Biological weed control methods offer sustainable alternatives to herbicides in globe artichoke cultivation. Effective strategies include:

Mulching: Applying natural materials like straw or wood chips around artichoke plants suppresses weed growth by blocking sunlight and preventing germination.

Cover cropping: Planting crops like clover or rye between artichoke rows competes with weeds for resources, reducing weed growth and enhancing soil health.

Manual weeding: Hand-pulling or hoeing weeds is effective in small-scale plantations. Regular inspection and removal prevent weed competition.

Biological agents: Introducing specific insects or pathogens that target weed species can control their populations. Expert guidance ensures no harm to artichoke plants.

Crop rotation: Alternating artichokes with non-host crops disrupts weed life cycles and reduces weed pressure.

Chemical herbicides include pre-emergent herbicides which are employed in the cultivation of both annual and perennial artichokes to control weeds. Directed applications of post-emergent herbicides are used to manage winter weeds specifically in perennial artichoke fields. Prior to weed emergence, diuron or simazine are applied to effectively control weeds. To tackle broad-leaved grassy weeds, it is recommended to apply Diuron (1.5kg ha^{-1}) + Trifluralin (1.2kg ha^{-1}) before planting. These herbicides have proven to be effective in ensuring effective weed control in globe artichoke cultivation. It's important to consider weed species, environmental factors, and farming practices when implementing these methods. Integrating multiple strategies and consistent monitoring ensure successful weed management in globe artichoke cultivation.

Harvesting

Approximately 60% of the world's globe artichoke production is focused on supplying the fresh market. The average weight of the heads ranges from 200 to 500 gm upto 700 gm, although smaller heads weighing 150 to 200 gm are also prevalent. These heads are typically sold with a peduncle measuring 4 to 6 cms in length (Macua, 2007). Additionally, the agro-industry, which includes canning and freezing, demonstrates considerable interest in cultivating this crop. (Macua *et al.*, 2011).

The method of propagating artichokes using suckers or off-shoots is a commonly employed technique and results in an earlier harvest, typically achieved within

6-7 months as compared to seed-sown crops (around 8 months). The harvest season usually begins in September-October. In seeded artichoke fields, the lack of uniform plant maturity and flowering extends the harvesting period (Schrader, 1994).

The optimal time for harvesting is when the buds have attained their maximum size, but prior to the bracts opening or the central fibrous portion of the bud (pappas) extending beyond the top edges of the receptacle. The largest bud, known as the terminal or primary bud, is harvested first, followed by the secondary and tertiary buds as they reach their full size. A single native plant can produce 12 or more stems, resulting in 40-50 edible buds that can be continuously harvested throughout the winter. On average, the yield per hectare is around 100-120 quintals.

The harvest of globe artichokes is typically done by hand, with a frequency of once or twice a week, adjusted according to weather conditions. In colder winter weather, harvest intervals for perennial artichoke fields may extend to 2 weeks or longer. Perennial artichoke plants typically undergo harvests of approximately thirty times or more over the course of the growing season.

Post-Harvest Management

The utilization of globe artichoke heads can vary based on the level of production and culinary customs in different nations. While fresh consumption is the most common, globe artichokes can also be preserved through freezing or canning methods (Costabile *et al.*, 2010; Lattanzio *et al.*, 2009). Globe artichoke being a perishable commodity, it is crucial to cool them as soon as possible after harvesting inorder to ensure the heads of globe artichokes maintain their high quality throughout the marketing process. According to Bratsch (2014), globe artichokes can be stored for a period of 3 to 4 weeks by maintaining a temperature range of 0-1°C and maintain a relative humidity level of 90-95%. Precooling methods, such as hydro cooling, are commonly employed to lower the temperature of globe artichokes below 5°C. However, room-cooling can also be considered as an alternative option, as stated by De Vos (1992). In the United States, grading of heads are typically sorted and classified based on their size and quality. They are then packed in field locations using waxed fiberboard cartons, as noted by Smith *et al.* (2008). While globe artichokes have limited storage capacity despite pre-cooling and cold storage techniques, notable advancements have been achieved through the utilization of propylene films, modified atmosphere packaging, and oxalic acid, as highlighted by Gil-Izquierdo *et al.* (2001; 2002), Alexopoulos *et al.* (2003), Leroy *et al.* (2010), and Ruiz-Jiménez *et al.* (2014). The research conducted by Restuccia *et al.* (2014) revealed that water ozonation and ozone enrichment in the atmosphere of the storage chamber can lead to a substantial reduction in pathogenic microorganisms. Furthermore, the application of ozonized water as a pre-treatment for globe artichokes, followed by storage in cooling chambers enriched with ozone, has been found to enhance the levels of total polyphenols and antioxidant activity in specific cultivars, such as Violet de Provence, as observed in the research (Lombardo *et al.,* 2015). To prolong the shelf life and maintain the quality of globe artichoke heads, it is essential to employ innovative techniques that can reduce respiration, prevent postharvest pathogen contamination, and minimize microbial spoilage.

Pests and Their Management

Globe artichoke cultivation can be negatively impacted by a range of pests, including arthropods and mollusks, which can lead to reduced yields and compromised quality. In Chile, a study conducted by Prado (1991) documented the presence of 22 pest species affecting artichoke crops. These pests include noctuids, which encompass loopers and cutworms, as well as aphids, leafminers, thrips, scarabs, snails, and slugs.

Cutworms, Caterpillars and Loopers (Lepidoptera: Noctuidae): Artichoke buds, leaves, stems, and related weeds serve as deposition sites for eggs by adult female insects. Larvae then feed on foliage and inflorescences, causing damage. The most significant impact is seen when larvae consume young shoots, hindering crop establishment. Established plants may suffer from larval feeding on inflorescence bracts, leading to distorted growth. Larvae can also contaminate developed inflorescences, reducing crop value due to worm presence and feces.

Control: Applying methomyl sprays before the initial three harvests reduces caterpillar damage. Combining intensive weed control with insecticide sprays like methomyl can reduce insecticide applications for managing noctuid species in artichoke crops with two hoe weeding controls in traditional management system.

Artichoke plume moth: The artichoke plume moth is the predominant pest found in artichoke plants. Its larvae, characterized by a yellowish coloration and a black head, feed on the leaves, stem, and developing buds of the plant.

Control: Yellow Sticky Traps and Neemoil can also be used for plume moth. Spraying of deltamethrin (0.0025 %) can be done for effective control of moths.

Aphids (Hemiptera: Aphididae): The presence of artichoke aphids (*C. elaeagni*), artichoke root aphids (*D. cynarae*), and bean aphids (*A. fabae*) becomes problematic when their colonies exceed 30% leaf coverage or infest the inflorescence. The presence of high aphid populations has detrimental effects on crop growth, leading to symptoms such as yellowing, leaf curling, reduced plant growth, and inflorescence colonization. Additionally, aphids secrete honeydew, promoting the growth of sooty mold fungi, which blackens foliage and inflorescences, inhibiting photosynthesis and leading to yield and quality losses.

Control: Biological control of aphids can be achieved through the use of generalist predators like lacewings, ladybird beetles, and hoverfly larvae which feed on aphids. Alternatively, aphids can be chemically controlled by applying a spray of malathion at a concentration of 0.05%.

Leaf miners (Diptera: Agromicidae): Damage caused by fly larvae, like the artichoke plume moth, mainly targets the foliage by consuming the parenchymal tissue beneath the leaf surface. Significant losses resulting from leafminer damage commonly occur when insect populations arrive at extremely high levels on comparatively small artichoke plants.

Control: Leaf miners can be trapped using Yellow Sticky Traps. Neem oil can also be utilized for control, where a spray is made by mixing 2 tablespoons of neem oil with a gallon of water, shaking well, and spraying it on the affected parts.

Thrips (Thysanoptera: Thripidae): This thrips species, like others, directly damages host plants by feeding on their tissues. The damage becomes evident as silver streaks or spots appearing on the leaves or bracts. Extensive damage to bracts can result in significant economic losses by diminishing the commercial value of inflorescences. Additionally, thrips have the tendency to deposit eggs directly into plant tissue, further exacerbating the harm caused.

Control: Debris and dead leaves should be cleared away to eliminate insect hiding spots. Floating row covers should be used as a protective barrier to prevent the plants from being accessed by thrips.

Mites (Acari: Tetranychidae): Artichoke plants are adversely affected by spider mites, as they pierce plant cells and extract sap, causing harm. The initial indications of feeding include the appearance of white dots on the leaves, later progressing to yellowing between veins. Severe infestations weaken plants, reduce photosynthesis, and can affect buds and flower heads.

Control: Controlling mites in globe artichoke involves cultural practices (hygiene, pruning, and removing infested parts), introducing predatory mites or beneficial insects, and, if needed, using acaricides.

The proba bug (Hemiptera: Miridae): *Proba californica* is a recent serious pest in central coast production, resembling the lygus bug (*Lygus hesperus*) in its life cycle and feeding habits. Both nymphs and adults feed on the juvenile leaves, injecting a toxin that stunts plant growth. Feeding on developing buds results in unsightly scars and deformities due to the phytotoxin.

Control: White sticky traps of 10 sq inch coated with petroleum jelly can be used to trap the bugs. Also, botanical insecticide like Sabadilla can be used to control the bugs.

Snails and Slugs (Stylommatophora: Helicidae and Limacidae): Artichoke leaves and stems are subject to damage from slugs and snails, as they graze on them, scraping the outer surface. Abundant populations of these mollusks can have a substantial impact on the yield and quality of the crop. In older plantations, there is a higher occurrence of mollusk populations, whereas annual plantings exhibit lower population levels.

Control: Debris and dead leaves should be removed to eliminate hiding places for snails and slugs. Barriers, such as copper tape or crushed eggshells, can be used to prevent their entry into specific areas.

Field mice and gophers (Rodentia: Cricetidae and Geomyidae): *Microtus spp* and *Thomomys bottae* can inflict significant financial harm in perennial globe artichoke fields. These pests consume the tender shoots, fleshy roots, and emerging buds of the artichoke plants, causing considerable damage.

Control: Trapping and baits control pests by attracting and capturing them to eliminate pest populations. Traps confine pests, while baits lure pests with toxic substances.

Diseases and Their Management

***Verticillium* wilt**: It is caused by the soil-borne fungi *Verticillium dahlia* Kleb., and it is a significant challenge in global artichoke production. Symptoms include wilting, chlorosis, stunted growth, and leaf necrosis. Vascular discoloration may not always be visible. Leaves initially turn yellow, then wilt and necrotize, often with a curved shape. Young leaves exhibit chlorosis and upward rolling. Defoliation is rare in wilt-affected artichokes.

Control: Effective strategies for managing Verticillium wilt in artichokes include the following disease control measures:

i) Selection of planting material free from any pathogen,

ii) Carefully selecting planting sites to bypass hazardous soils.

iii) Minimizing *Verticillium dahlia* inoculum in the soil.

iv) Protecting uninfected planting material from soil-borne pathogens.

v) Using resistant varieties.

Botrytis blight (*Botrytis cinerea*): When weather conditions turn warm and moist, fungal growth and decay become evident on tissue that has suffered damage from insects or frost. This fungus can infect both the plant and the flower bud.

Control: Removal of infected buds and timely planting to reduces frost injury.

Powdery mildew (*Leveillula taurica*): The fungal diseases of powdery mildew (*Leveillula taurica*) can result in significant financial losses in artichoke cultivation. These pathogens target bracts and foliage, leading to early leaf aging, leaf shedding, and diminished economic value of affected buds.

Control: Effective management involves maintaining cleanliness, controlling irrigation to minimize leaf moisture, considering fungicide applications, exploring resistant varieties, and utilizing biological control methods.

Bacterial disease

Bacterial crown rot (*Erwinia chrysanthemi*): In warm weather, artichoke plants are susceptible to stunted growth and wilting caused by this condition. During the advanced stages, plants may suffer from collapsing as the root and crown tissues become soft, rotting and acquiring a black or brown discoloration. Disease transmission via harvesting tools is suspected.

Control: The primary method of controlling the disease is to utilize uncontaminated propagation material and prevent disease spread during harvesting and propagation through the use of sterilized tools.

Viruses

Curly dwarf disease: Artichoke curly dwarf virus (ACDV), a type of Potexvirus, causes disease in globe artichokes (*Cynara cardunculus* sub. scolymus L.). Symptoms include curling and dwarfing, resulting in irregular bud production, reduced vigor, and potential plant death. It primarily affects artichoke, cardoon, zinnia, and milk thistle (*Silybum marianum*).

Control: Control measures include the use of virus-free plant stocks when establishing new plantations. The successful production of virus-free regenerants of the Brindisino cultivar, maintaining its early flowering and characteristic head shape, has been accomplished through the effective implementation of a combined approach involving meristem-tip culture and thermotherapy technique.

Future Prospect

The future prospects for crop improvement of Globe Artichoke are promising. Research efforts are focused on developing improved varieties with enhanced yield potential, disease resistance, and adaptability to different environmental conditions. Genetic studies and breeding programs aim to incorporate desirable traits such as reduced spines, improved uniformity, and increased tolerance to abiotic stresses. Additionally, advancements in cultural practices, such as optimized irrigation and nutrient management, can contribute to the overall productivity and sustainability of Globe Artichoke cultivation. These advancements in crop improvement hold great potential for ensuring the continued growth and success of Globe Artichoke as an important agricultural crop.

References

Alexopoulos, A. A., Akoumianakis, K. A., & Passam, H. C. (2003). The storage of globe artichokes under modified atmospheres.

Bedini, L., Lucchesini, M., Bertozzi, F., & Graifenberg, A. (2012). Plant tissue cultures from four Tuscan globe artichoke cultivars. *Central European Journal of Biology*, 7: 680-689. https://doi.org/10.2478/s11535-012-0064-x

Bianco, V. V. (2000). Present situation and future potential of artichoke in the Mediterranean basin. In IV International Congress on Artichoke 681 (pp.39-58). https://doi.org/10.17660/ActaHortic.2005.681.2

Bonasia, A., Conversa, G., Lazzizera, C., Gambacorta, G., & Elia, A. (2010). Morphological and qualitative characterisation of globe artichoke head from new seedpropagated cultivars. *Journal of the Science of Food and Agriculture*, 90(15), 2689-2693. https://doi.org/10.1002/jsfa.4141

Bratsch, T. (2005). Speciality crop profile: Globe artichoke.

Cadinu, M., Repetto, A., Frau, A., Beneventi, S., & Meloni, S. (2004). Influence of the explant type on the phenotypic changes in micropropagated plants of artichoke. *Acta Hortic.* 660:373-80.

Campanelli, A., Ruta, C., Tagarelli, A., Morone-Fortunato, I., & De Mastro, G. (2014). Effectiveness of mycorrhizal fungi on globe artichoke (*Cynara cardunculus* L. var. scolymus) micropropagation. *Journal of Plant Interactions*. **9**(1): 100-106 https://doi.org/10.1080/17429145.2013.770928

Castiglione, V., Cavallaro, V., Di Silvestro, I., & Barbera, A. C. (2007). Acclimatization of micro propagated globe artichoke (*Cynara cardunculus* L. subsp. scolymus (L.) Hegi) plantlets as affected by mycorrhizal inoculum, transplantation time and genotype. In III International Symposium on Acclimatization and Establishment of Micropropagated Plants 812 (pp. 461466). https://doi.org/10.17660/ActaHortic.2009.812.66

Ciccarese, F., Crinò, P., Raccuia, S. A., & Temperini, A. (2009, June). Use of resistant cardoons as rootstocks for the control of Verticillium wilt in globe artichoke. In VII International Symposium on Artichoke, Cardoon and Their Wild Relatives 942 (pp. 201-205). https://doi.org/10.17660/ActaHortic.2012.942.26

Costabile, A., Kolida, S., Klinder, A., Gietl, E., Bäuerlein, M., Frohberg, C., & Gibson, G. R. (2010). A double-blind, placebo-controlled, cross-over study to establish the bifidogenic effect of a very-long-chain inulin extracted from globe artichoke (*Cynara scolymus*) in healthy human subjects. *British journal of nutrition*, **104**(7)**:** 10071017.

De Nardi, F. S., Calvete, E. O., Reolon-Costa, A., Costa, R. C., Cravero, V. P., Scheffer-Basso, S. M., & Chiomento, J. L. T. (2015, September). Morpho-agronomic variability of artichoke seed-propagated hybrids. In IX International Symposium on Artichoke, Cardoon and Their Wild Relatives 1147 (pp. 6976).https://doi.org/10.17660/ActaHortic.2016.1147.10

De Pascale, S., Rouphael, Y., Graziani, G., Ritieni, A., Colla, G., Fiorillo, A., & Saccardo, F. (2015, September). Yield, mineral and polyphenolic composition of two seed-propagated globe artichoke cultivars. In IX International Symposium on Artichoke, Cardoon and Their Wild Relatives 1147 (pp. 8388). https://doi.org/10.17660/ActaHortic.2016.1147.12

De Vos, N. E. (1992). Artichoke production in California. *HortTechnology*, **2**(4)**:**438-444.

Deligios, P. A., Tiloca, M. T., Sulas, L., Buffa, M., Caraffini, S., Doro, L.,& Ledda, L. (2017). Stable nutrient flows in sustainable and alternative cropping systems of globe artichoke. *Agronomy for Sustainable Development*,37:112.https://hdl.handle.net/11566/286524

Di Venere, D., Pieralice, M., Linsalata, V., Gatto, M. A., Sergio, L., & Calabrese, N. (2015). Biochemical evaluation of artichoke cultivars propagated by seed. In IX International Symposium on Artichoke, Cardoon and Their Wild Relatives 1147 (pp. 89-94). https://doi.org/10.17660/ActaHortic.2016.1147.13

El Boullani, R., Lagram, K., El Mousadik, A., & Serghini, M. A. (2017). Effect of explants density and size on the in vitro proliferation and growth of separated shoots of globe artichoke (*Cynara cardunculus* var. scolymus L.). *J. Mater. Environ. Sci*, **8**(7): 2469-73.

Feleafel, M. N. (2005). Effects of decapitation time and nitrogen rate on vegetative growth, yield and quality characteristics of globe artichoke. *J. Agric. Environ. Sci. Alex. Univ. Egypt*, **4**(2)**:** 78-95.

Gil-Izquierdo, A., Conesa, M., Ferreres, F., & Gil, M. (2002). Influence of modified atmosphere packaging on quality, vitamin C and phenolic content of artichokes (*Cynara scolymus* L.). *European Food Research and Technology*, **215:** 21-27.

Gil-Izquierdo, A., Gil, M. I., Conesa, M. A., & Ferreres, F. (2001). The effect of storage temperatures on vitamin C and phenolics content of artichoke (*Cynara scolymus* L.) heads. *Innovative Food Science & Emerging Technologies*, **2**(3)**:**199202. https://doi.org/10.1016/S1466-8564(01)00018-2

Grando, M. F., Augustin, L., Suzin, M., Calvete, E. O., Comin, R. C., Costa, A. R., ... & Donida, B. (2011). Micropropagation of globe artichoke'Nobre-UPF', a Brazilian cultivar used for industrial purpose. *Acta horticulturae*, (923), 147-154.

Halter, L., Habegger, R., & Schnitzler, W. H. (2000). Field technologies for commercial production of artichoke (*Cynara scolymus* L.) in Germany. In IV International Congress on Artichoke 681 (pp. 169174). https://doi.org/10.17660/ActaHortic.2005.681.18

Iapichino, G. (2013). Micropropagation of Globe Artichoke (*Cynara cardunculus* L. var. scolymus). *Protocols for Micropropagation of Selected Economically Important Horticultural Plants*, 369-380.

Ierna, A., &Mauromicale, G. (2012, April). Effects of organo-mineral fertilizers on yield characteristics of globe artichoke. In VIII International Symposium on Artichoke, Cardoon and their Wild Relatives **983** (pp. 275-281).

Lattanzio, V., Kroon, P. A., Linsalata, V., &Cardinali, A. (2009). Globe artichoke: A functional food and source of nutraceutical ingredients. *Journal of Functional Foods*, **1**(2)**:**131-144. https://doi.org/10.1016/j.jff.2009.01.002

LENZI, A., & BALDI, A. (2015). Artichoke (*Cynara scolymus* L.) as cash-cover crop in an organic vegetable system. *ActaagriculturaeSlovenica*, **105**(1)**:**5360.

Leroy, G., Grongnet, J. F., Mabeau, S., Corre, D. L., &BatyJulien, C. (2010). Changes in inulin and soluble sugar concentration in artichokes (Cynara scolymus L.) during storage. *Journal of the Science of Food and Agriculture*, **90**(7)**:** 1203-1209.

Lombardo, S., Pandino, G., Restuccia, C., Muratore, G., Licciardello, F., Mauro, R. P., ... &Mauromicale, G. (2015). Effect of cultivar x ozone treatment interaction on the total polyphenols content and antioxidant activity of globe artichoke. *Italian Journal of Agronomy*. **10**(2)**:** 105-107.

Lombardo, S., Restuccia, C., Muratore, G., Barbagallo, R. N., Licciardello, F., Pandino, G., Scifò, G.O., Mazzaglia, A., Ragonese, F and Mauromicale, G. (2017). Effect of nitrogen fertilisation on the overall quality of minimally processed globe artichoke heads. *Journal of the Science of Food and Agriculture*, **97**(2): 650-658

Macua, J. I. (2006). New horizons for artichoke cultivation. In VI International Symposium on Artichoke, Cardoon and Their Wild Relatives 730 (pp. 39-48).

Macua, J. I., Lahoz, I., & Bozal, J. M. (2009, June). Evolution in seed-propagated artichoke growing in cold zones of Spain. In *VII International Symposium on Artichoke, Cardoon and Their Wild Relatives* **942** (pp. 331337)

Montervino, C., Lotti, C., Elia, A., & Conversa, G. (2006). Micropropagation of the early artichoke cultivar'Violet du Provence'. In VI International Symposium on Artichoke,Cardoon and Their Wild Relatives 730 (pp.127134.

Morello, N., Santoiemma, G., Ierna, A., & Mauromicale, G. (2000). Improvement of″ ovoli′ production in globe artichoke by removal of the epigeal part of the plant. In IV International Congress on Artichoke 681 (pp. 251-256).

Pagnotta, M. A., Rey, N. A., Mondini, L., Aringoli, R., Jordan, R., & Saccardo, F. (2015, September). Assessment of artichoke hybrids under USA and Italian conditions and the hereditability of some important traits. In IX International Symposium on Artichoke, Cardoon and Their Wild Relatives 1147 (pp. 257264

Pandino, G., Lombardo, S., Antonino, L. M., Ruta, C., & Mauromicale, G. (2017). In vitro micropropagation and mycorrhizal treatment influences the polyphenols content profile of globe artichoke under field conditions. *Food Research International*, 99, 385-392.

Pandozy, G., Trinchera, A., Rinaldi, S., Rea, E., Saccardo3, F., & Crinò, P. (2015, September). Grafting of artichoke for the sustainable management of Verticillium wilt. In IX International Symposium on Artichoke, Cardoon and Their Wild Relatives 1147 (pp. 265-270

Restuccia, C., Lombardo, S., Pandino, G., Licciardello, F., Muratore, G., &Mauromicale, G. (2014). An innovative combined water ozonisation/O3-atmosphere storage for preserving the overall quality of two globe artichoke cultivars. *Innovative Food Science & Emerging Technologies*, **21**:82-89

Rey, N. A., Jordan, R., Saccardo, F., & Pagnotta, M. A. (2015). A successful strategy to obtain artichoke hybrids. In IX International Symposium on Artichoke, Cardoon and Their Wild Relatives 1147(pp. 357-368).

Riahi, J., Nicoletto, C., Bouzaein, G., Sambo, P., & Khalfallah, K. K. (2017). Effect of vegetative propagation materials on globe artichoke production in semi-arid developing countries: agronomic, marketable and qualitative traits. *Agronomy*, **7**(4): 65.

Ruíz-Jiménez, J. M., Zapata, P. J., Serrano, M., Valero, D., Martínez-Romero, D., Castillo, S., & Guillén, F. (2014). Effect of oxalic acid on quality attributes of artichokes stored at ambient temperature. *Postharvest Biology and Technology*, **95:**

Ruta, C., Tagarelli, A., Vancini, C., & De Mastro, G. (2015). Evaluation of commercial arbuscular mycorrhizal inoculants on micropropagated early globe artichoke during the acclimatization stage. In IX International Symposium on Artichoke, Cardoon and Their Wild Relatives 1147 (pp.369374)

Ryder, E. J., De Vos, N. E., & Bari, M. A. (1983). The globe artichoke (*Cynara scolymus* L.). *HortScience,* **18**(5): 646653.

Smith, R., Baameur, A., Bari, M., Cahn, M., Giraud, D., Natwick, E., & Takele, E. (2008). Artichoke production in California. 10.3733/ucanr.7221

Sonnante, G., Pignone, D., and Hammer, K. (2007). The domestication of artichoke and cardoon: from Roman times to the genomic age. Annals of Botany, **100**(5): 1095-1100.

Spanu, E., Deligios, P. A., Azara, E., Delogu, G., &Ledda, L. (2018). Effects of alternative cropping systems on globe artichoke qualitative traits. *Journal of the Science of Food and Agriculture*, **98**(3):1079-1087.

Tavazza, R., Rey, N. A., & Pagnotta, M. A. (2015). Globe artichoke in vitro conservation protocol to meet germplasm preservation and production management. In IX International Symposium on Artichoke, Cardoon and Their Wild Relatives 1147 (pp. 421-428).

Temperini, O., Calabrese, N., Temperini, A., Rouphael, Y., Tesi, R., Lenzi, A., ... & Colla, G. (2013). Grafting artichoke onto cardoon rootstocks: Graft compatibility, yield and Verticillium wilt incidence. *Scientia Horticulturae*, 149, 22-27.

Trinchera, A., Pandozy, G., Rinaldi, S., Crinò, P., Temperini, O., & Rea, E. (2013). Graft union formation in artichoke grafting onto wild and cultivated cardoon: *An anatomical study. Journal of Plant Physiology*, 170(18), 1569-1578.

Virdis, A., Motzo, R., & Giunta, F. (2014). The phenology of seedpropagated globe artichoke. *Annals of applied biology*, 164(1), 128-137.

Welbaum, G. E. (1994). Annual culture of globe artichoke from seed in Virginia. *Hort Technology*, **4**(2): 147-150.

22

Advances in Production Technology of Rhubarb

***Dhaneshvari Arya*[*1], *Prakash*[2], *Nitin Yadav*[3] & *Rameshwar Jangu*[4]**

[1&2]*Ph.D. Research Scholar, Division of Vegetable Science, Indian Agricultural Research Institute, New Delhi, India-110012*

[3]*Ph.D. Research Scholar, Department of Horticulture, Banaras Hindu University, Varanasi, Uttar Pradesh, India-221005*

[4]*Ph.D. Research Scholar, Department of Vegetable Science, Punjab Agricultural University, Ludhiana, Punjab, India-141004*

[*]*Email: Arya081998@gmail.com*

Introduction

The "rhubarb and pie plant," also known as *Rheum rhaponticum* L., is a member of the Polygonaceae family. Diploid (2n = 2x = 22), tetraploid (2n = 4x = 44), or hexaploid (2n = 6x = 66) rheum species are the three possible forms. The long, thick leafstalk, often called the petiole, is used to make pies and sauces. The rhubarb stalks' colour may range from the typically recognized crimson red to pale pink with specks to plain light green. Rhubarb is a perennial plant that thrives in cold climates and is drought and winter tolerant. Edible petioles and leaves are the first springtime shoots to emerge. rhubarb was first cultivated in Italy and introduced to the continent in the year 1608. The earliest recorded use of rhubarb in the United States was in 1778, but by 1806 it had achieved broad acceptance. According to Lindley's Treasury of Botany, the name comes from the Greek word rheo, meaning "to make one vomit," because of the root's laxative properties. It is a good source of carbohydrates (3.8g), protein (0.74 g), fibre (1.75 g) and fat (0.13g) and minerals like calcium (130 mg) and potassium (360 mg) per 100 grams of edible portion. There are numerous rhubarb cultivars contain bioactive substances such as flavonoids, anthraquinones, glycosides, tannins, volatile oils, and saponins. The chemical composition of rhubarb includes anthraquinones, anthrones, stilbenes, tannins, polysaccharides, etc., which show extensive pharmacological activities including gastrointestinal regulation, anticancer and antimicrobial properties, hepatoprotection, cardiovascular and cerebrovascular anti-inflammatory protection, etc. The primary qualities of rhubarb quality are taste and aroma, which are influenced by its chemical components.

Botany

Rhubarb, one of the ancient and most valued plants, has deep roots, hollow stalks, and clusters of small white, green, or purple-red flowers. Panicles with small, bluish-greenish-white flowers are formed; they occur successively. Flowers are not capable of self-pollination since they are protandrous. Utilizing pollen from higher blooms to pollinate lower ones. It is substituted for fruits in the diet.

Taxonomy

There are more than 60 species of the genus *Rheum* (rhubarb, Polygonaceae) that are used as food or medicine throughout Asia, Europe, and other parts of the globe. While the petioles of garden rhubarbs (*R. rhabarbarum*/*R. undulatum*) and rhapontic rhubarbs (*R. rhaponticum*) are consumed for their nutritional value, the roots and stems of these plants are also utilized for medicinal purposes. *R. rhaponticum*, *R. rhabarbarum* and other rhubarbs have been the subject of thorough reviews of the current understanding of their phytochemical profile, ethnomedicinal significance, and principal lines of physiological action. Therefore, these concerns have only been glossed over in this study. Both species are used to treat a variety of gastrointestinal complaints in traditional medicine. Additionally, *R. rhaponticum*-based combinations were utilized to treat cardiovascular disease, lung illnesses, and reproductive system issues including uterine and breast aches. Purgative and sedative *R. rhabarbarum* was given to patients. Both *R. officinale* Baill. (Chinese rhubarb) and *R. palmatum* L. (palm rhubarb) are well-known medicinal herbs used to treat gastrointestinal problems, chronic renal failure inflammations, and liver ailments. In both traditional Chinese and Ayurvedic medicine, *Rheum emodi* Wall ex. *Meissn* is used to treat a wide range of conditions, including cancer, ulcers, headaches, diarrhoea, and liver diseases. Both *R. officinale* and *R. ribes* L. (Syrian rhubarb) are used to treat obesity in traditional Turkish medicine. In China, the rhizomes of *Rheum palmatum* L., *Rheum tanguticum* Maxim. ex Balf., and *Rheum officinale* Baill are used for medicinal purposes, whereas the petioles of *Rheum rhaponticum* L. are consumed as a vegetable in Europe.

Soil and Climate

Rhubarb can endure cold temperatures. Warm winters and dry summers cause the plant to hibernate in the summer, while milder months and early spring are when it thrives. Plants hibernate until the warmer months return when the temperature falls below freezing in the winter. 3.3 °C is the bare minimum temperature at which plant vegetative components would be lost. When the summertime temperature is above 23.9°C and the wintertime temperature is below 4.5°C, the plant likewise does not grow well. If the crop is irrigated and the weather is ideal, it appears that a dormant period is not essential because the plant may continue to develop all year long. A soil that is both organic matter-rich and well-drained is necessary for growing rhubarb well. While a little acidic soil may let it flourish, neutral soil is best for its growth.

Field Preparation

A well-prepared soil not only offers enough space for tuber growth but also aids in soil moisture retention. Clods are broken and the farmland is ploughed to a depth of approximately 20 to 35 cm. To control the issues of soil-borne diseases and perennial weeds, hot weather cultivation in the plains during the summer months of April to June involves deeply ploughing the ground and turning the soil one or two times. The fields are adequately levelled and have appropriate drainage installed once the green manure has been buried. Fields are prepared for planting by being ploughed with mould board ploughs or disc harrows, then tilled once or twice with a tiller.

Commercial Varieties

Red Types: Crimson Cherry, Crimson Red, and Crimson Wine are further names for the crimson variety. Under typical temperature and moisture conditions, Crimson develops vividly coloured red stalks with the distinctive characteristic of being crimson all throughout. Valentine, Canada Red, and Cherry Red are further robust red variants. Californian growers of Cherry Red generate tall, thick, deep-red stalks.

Speckled Types (pink): Victoria provides excellent-quality, big stalks. It has smooth ribs and is long and rounded. It sprouts pink speckling on a stem that is pale green. The pink hue is more prominent towards the base of the stalk and gradually fades to a solid green tint at the top. Victoria is frequently employed in forcing. Victoria and Strawberry are quite similar, and they could even be the same kind. Another productive pink variety is MacDonald. German Wine resembles Victoria, but it is more active and intensely coloured, generally with deeper pink speckling on a stalk of green.

Green Types: The robust, cold-resistant Riverside Giant has a huge diameter, long, green stems.

Seed rate and Seed treatment

Rhubarb is not propagated from seed since seedlings are not true to type. It required 10,000 divisions for the planting of one ha. Land.

In the summer when divisions are to be produced, identify healthy, disease-free plants and stake them. Plug the selected seedlings in the autumn or early spring when they are dormant. Store in a frigid location and guard against repeated freezing and thawing, drying out, and harm to buds. Cut each plant into as many divisions as you can just before planting time in the spring, each with one or two well-developed buds and as much of the nearby root material as feasible. Prevent drying out of divisions both before and during planting. Depending on the age and cultivar, each parental plant can produce a division between 4 and 20 clones.

Sowing

Typically, crowns generated during the following years are divided to proliferate rhubarb. In the late fall or early spring, crowns are separated. Every four years, plants need to be split and replanted to maintain the bed healthy. If plants are not divided, they may grow huge and produce an excessive number of stalks. As soon as they emerge, seed stalks must be cut off. New types are created by seed propagation. In the spring, when cold temperatures are frequent, the crowns are planted. In October, when the weather is not below freezing, the crowns are planted. Growing rhubarb requires good drainage. Rows 120 cm apart with plants 120 cm apart are often recommended spacing for commercial rhubarb growing. In advance of planting, 15 cm deep trenches are excavated. Crown pieces are put in the trench 5 cm below the soil's surface. The earth is firmed by pulling it over it.

Nutrition Management

Based on recommendations from a soil test, all lime additions, fertilizer applications, and manure applications should be made. When test results are unavailable, the general standards listed in tables 1 through 4 of this guide's section on soil and water management may be followed.

In a very fertile soil that is rich in organic matter and well-drained, rhubarb grows nicely. The year before planting, it is preferable to have a green manure crop. 45 tonnes of manure per acre may also be applied in the fall. (It would be advantageous to apply manure every fall.) Applying lime will help keep the pH of the soil between 6.0 and 6.8. (Rhubarb will tolerate soil acidity as low as 5.0 but yields and fertilizer efficiency will suffer). The calcium level should be 3000 kg/ha for all crops in order to comply with the most recent soil testing techniques. Rhubarb has a high need for the nutrient nitrogen. Add only as needed throughout the establishing year; if not, apply nitrogen at bud break together with the appropriate amounts of phosphate and potash. Apply one or two side dressings of nitrogen to crops intended for forcing or after harvest (wait until strong regrowth has started). Manure applications will help this crop meet some of its nitrogen needs. Phosphorus is crucial during the first year of establishment. Rhubarb should be grown in phosphorus-rich fields. Potash is only applied if the soil levels are insufficient, although it is frequently necessary if high yields are being harvested without the use of manure. On sandy soils with little organic matter, magnesium and/or Sulphur may be necessary. Gypsum and dolomitic limestone usage, as well as foliar sprays of Epsom salts, may be taken into consideration. Boron is one of the micronutrients that may be important for the health of buds and roots. It may be sprayed on leaves or soaked into the soil when used as fertilizer. Apply genuine boron at a rate of 1 to 2 kg per acre per year Avoid burning any leaves or roots during the establishing year. N, P, and K should be broadcast in the spring, and manure should be used in the fall. Apply nitrogen through side dressing, broadcasting, or using an irrigation system. The broadcast fertilizer will be mixed into the soil and certain weeds will be controlled by cultivation between the rows. Fertilizers and manuring Before ploughing, poor organic matter soil should receive 50 tonnes/ha of well-rotted organic manure. It requires an addition of 100–150 kg of each NPK. Just prior to planting at the bottom of the furrows during the first year, the soil is amended with fertilizers.

Irrigation Management

After the harvest season, keep the soil wet enough to promote healthy regrowth. Weather allowing, a second crop may be collected in early to mid-July. Thereafter, the field may be replanted and left to fall dormant in late July and early August. Growth will be hindered the next spring if harvested in July. Although the type of soil affects the frequency of water application rather than the overall amount of water required, Lighter soils require less water applied each application but more frequent watering. It required 427 mm/year of water for better growth and yield.

Weed Management

A mulch of various sorts can be used in the rows and between the plants (most commonly straw). Weeds can be controlled by frequent shallow cultivation before sprouting. To control grass weeds, apply gramoxone or paraquat at 2.25-3.75 l/ha, or glyphosate or round-up at 2-3 kg a.i./ha.

Physiological Disorders and Their Management

When exposed to moisture, petioles lacking small leaf lamina can break. Petioles that have grown too long get sour. Petioles may become unsightly when they are damaged by sand or excessive treatment.

Management

- Deep ploughing the field, adding organic matter to the soil, and watering the field are ways to control this physiological disorder.
- Maintain proper moisture in the field.
- Adopt the techniques suggested in the recommendation package.
- Minimize the excessive chemical treatment.

Major Insect-Pests and their management

1. Insects Potato Stem Borer

Characteristics: The potato stem borer is a caterpillar that is about 3.5 cm long and pinkish-white in colour when fully mature. The insect's early stages only target weeds, particularly couch grass. Later, they go into plants like rhubarb that have thicker stems. They could migrate along the stalk, burrowing into the centre of each stem as they go. In August, mature moths deposit their eggs on grass stems. It takes until the spring after that for the eggs to hatch. In June and the first part of July, the damage is probable. A serious infestation may result in a crop that cannot be sold.

Control: When couch grass and other weeds are eliminated from the rhubarb plantation and its surroundings, this pest is not an issue. To reduce the field's appeal to the adult moth for egg laying, weeds should be controlled. This insect can be efficiently controlled by burning the afflicted fields or field borders in the early spring.

2. Tarnished plant bug

Characteristics: Tarnished plant bugs are roughly 5-6 mm long and range in colour from light brown to reddish brown. They happen all through the season. They move quickly and are quite active. They can hurt rhubarb by eating the tender leaves. They wilt the leaves and deform them by puncturing the stem with their mouthparts. New plants are mostly plagued by tarnished plant pests.

Control: Keep weeds out of surrounding areas and plantings. Don't plant close to legumes.

3. Slugs

Characteristics: Plantings with heavy soils, poor drainage, and weedy conditions may experience a slug problem. Slugs feed at night by rasping the stem's surface, causing unattractive scars that make the stem less marketable.

Control: Ensure adequate soil drainage. Take steps to manage weeds. When harvesting, clear the field of debris and leaves. In fields where slug damage is a possibility, avoid using mulch and/or manure.

4. Other Insect Pests

Black bean aphid: *Aphis fabae* has occasionally been a concern. The leaves may curl or wilt in large infestations. Aphids can transfer viruses as well.

Flea beetles: By chewing on the leaves, flea beetles can harm newly planted vegetation. They are most active when it is hot and dry outside.

Control: Exercise effective weed control. For either of these pests, there are no registered products.

Major Diseases and their management

1. **Diseases Leaf Spots** - Ascochyta and Ramularia (fungi) These fungi produce round or angular patches that can vary in size and have beige cores encircled by red areas. Affected tissue may drop off as it expires, producing obscenely big, ragged holes in the leaves. Infected propagation stock and plant waste both harbour fungi that overwinter.

   Control: After the first hard frost, remove and destroy the leaves. Remove stems with spotted leaves first while harvesting. Apply a fungicide for controlling Botrytis.

2. **Botrytis Rot (fungus) -** This may lead to forced rhubarb's leaf, stem, and crown rot. Where there is inadequate air circulation and heavy humidity, the disease worsens.

   Control: Maintain rigorous hygiene. At the first indication of the illness and every 7 days afterwards, apply the specified fungicide.

3. **Root and Crown Rots (fungi & bacteria)**

   Characteristics: Plants start to waste money. It's possible for leaves to transform from yellow to crimson before falling. When the crowns are sectioned, a brown-black deterioration is seen. Small feeder roots are lacking in big roots. Large brown-black lesions may be found on larger roots.

   Control: Invest in healthy propagation stock. Pick fields that have good drainage. Delete and eliminate sick plants.

4. **Virus**

   Characteristics: Rhubarb is reported to have several viruses. Turnip mosaic, arabis mosaic, and cherry leaf roll virus are thought to be the most prevalent according to assessments from British Columbia and the UK. These viruses induce mottling and ring spotting on leaves and have a wide range of hosts. They could be spread by contaminated planting stock.

   Control: Collect and sow wholesome seedlings. Keep virus-free crowns apart from crowns that have been exposed to viruses.

Yield

The average yield of rhubarb is 150-400 q/ha.

Harvesting and Post-Harvest Management

When rhubarb is planted, it is picked after the second year. The stalks are tugged during harvesting not cut and the leaf blades are subsequently removed. Every year, harvesting is possible for 8 to 10 weeks. For around 10–12 years, the plants may be kept in the field. The plants can be pulled from the field when they are extremely congested and the leaf stalks have shrunk. The stalks are then given a water wash before being tied into 0.5kg bunches for the neighbourhood market. For transportation, the gathered stalks are cleaned, bundled into bundles weighing at least 0.5 kg, and placed in cartons coated with paper. Grading ought to be done by stalk length, colour, and quality. It is gradable into grades I and II. Harvested stalks keep their quality for around 4 weeks when kept at a temperature of 1-2°C and 92% relative humidity. By wrapping in moisture-proof film and storing at 0°C and 90% relative humidity, the storage life can be further prolonged. The shelf life is doubled at 21°C after 2 minutes of hot water treatment at 52°C.

Modern Technology

Propagation

Rhubarb can be reproduced vegetatively by division of the mother plant or from plantlets derived from in vitro preparations, as well as generatively from seedlings grown in a nursery or greenhouse. Rhubarb plants that have been reproduced in this way are typically not particularly consistent and do not carry over parental features. The height, habit, form, and intensity of petiole colour, as well as the number of petioles per plant, vary amongst plants of the same cultivar. The practice of seed propagation is not advised in horticulture. On the other hand, the poor yield of mother plants and the potential for the spread of viruses, particularly viral illnesses, limit the production of rhubarb division of crowns. In the literature, it is frequently emphasized how popular producing rhubarb from micro propagated plants is become. This approach offers the chance to create highly vigorous, genetically stable plants that are also virus-free. The following field acclimation of plants in nurseries and adaptation to the environment, as well as the phenotypic, physiological, and biochemical changes that take place, are major issues with tissue culture plant propagation. After being transplanted to the field, in vitro propagated rhubarb plants exhibited substantial somaclonal diversity in terms of their morphological characteristics, including plant habit, petiole shape and colour, in addition to low winter hardiness's observation of micro propagated plants growing in the field revealed a high frequency of alterations. It has been demonstrated in studies evaluating cultivated progeny that plants can undergo morphological changes over time when grown in an open environment. Contrary to traditional propagation techniques, tissue culture conditions, namely the concentration and kind of growth regulators utilized, were a direct source of stress and led to physiological alterations in plants.

Forcing

In areas where the environment is conducive to the growth of robust crowns, the creation of rhubarb stalks during the winter is referred to as the "forcing of rhubarb." Three-year-old plants' roots are pulled out late in the fall and kept outside for two weeks until they are completely frozen. To stop rainwater from evaporating, a thin coating of earth is applied. The freezing is done to give the buds time to relax. Then, in hotbeds in specially designed forcing structures or in any other dark or semi-dark sites where moderate temperatures may be maintained, the frozen roots are placed. Since light is undesirable, the forcing structure's windows, if any, should be covered to block off the light. The buds can now begin to sprout. The stems take on a deep pink or red colour and high quality in dim or gloomy environments. The leaf blade does not expand much. Heat may be produced using hot water furnaces. The ideal temperature for healthy stalk growth would be 15.6°C. The colour and craftsmanship are both quite beautiful. Poor growth is expected at 10°C. The forcing structure's floor is coated with 5–7.5 cm of soil, and the crowns are arranged as close together as possible. To make simple stalk harvesting possible, a walking space can be provided every 1.5–2.0m. The crowns sprout and the stalks extend due to the high warmth. They may be picked, packaged, and sent to markets after they reach a length of 45 cm.

Cooking and Nutrition

Rhubarb is prepared by canning, freezing, and cooking. Oxalic acid is what gives it its acidic flavour. The acidic flavour is softer when sugar is added. Provitamin A and vitamin C are found in rhubarb. The oxalates found in rhubarb leaves make them toxic. Attempts to utilise the leaves as a vegetable green have resulted in fatalities. Only consume the petioles (stalks). Before using rhubarb in recipes, remove any remaining leaf tissue. In cookbooks and on the Internet, there are countless recipes for rhubarb bread, crisps, cobblers, pies, tarts, teas, and wines. For thoroughness, only one recipe is included here.

Future Prospects

The species and origin of the rhubarb used in the majority of these trials have not been reported, and the cases included in the assessments lack information on the participants' age, race, and other characteristics. To get an understanding, more scientific, rigorous, and thorough experimental studies are required. Improvement of more efficient kinds. Collection, conservation, and characterization of wild species for pest and disease resistance. In the future, research including the development of new technologies perfect for the stable, year-round, and inexpensive production of seedlings is necessary, so that these vegetables become more accessible to farmers worldwide.

References

Barna KS, Karihaloo JL and Tiwari S. (2007). Long-term field performance of micro-propagated Aloe vera plants. *Plant Cell Biotechnology and Molecular Biology*. 113-118.

Cao YJ, Pu ZJ, Tang YP, Shen J, Chen YY, Kang A, Zhou GS and Duan JA. (2017). Advances in bio-active constituents, pharmacology and clinical applications of rhubarb. *Chinese Medicine*. 12(1):1-12.

Cao YJ, Pu ZJ, Tang YP, Shen J, Chen YY, Kang A, Zhou GS and Duan JA. (2017). Advances in bioactive constituents, pharmacology and clinical applications of rhubarb. *Journal of Traditional Chinese Medicine*. 12:36.

Clapa D, Borsai O, Harţa M, Bonta V, Szabo K, Coman V and Bobiş O. (2020). Micropropagation, genetic fidelity and phenolic compound production of *Rheum rhabarbarum* L. *Plants*. 9(5):656.

Kalisz S, Oszmiański J, Kolniak-Ostek J, Grobelna A, Kieliszek M and Cendrowski A. (2020). Effect of a variety of polyphenols compounds and antioxidant properties of rhubarb (*Rheum rhabarbarum*). *Lwt*. 118:108775.

Kolodziejczyk-Czepas J and Liudvytska O. (2021). *Rheum rhaponticum* and *Rheum rhabarbarum*: A review of phytochemistry, biological activities and therapeutic potential. *Phytochemistry Reviews*. 20(3):589-607.

Kozak D. and Saaata A. (2011). Effects of cytokinins on in vitro multiplication of rhubarb (*Rheum rhaponticum* L.) 'Karpow Lipskiego' shoot and ex vitro acclimatization and growth. *Acta scientiarum polonorum hortorum cultus*. 10(4):75–87.

Lee MR, Hutcheon J, Dukan E and Milne I. (2017). Rhubarb (*Rheum* species): the role of Edinburgh in its cultivation and development. *Journal of the Royal College of Physicians of Edinburgh*. 47(1):102-109.

Ma, X. and Gang, D. R. (2006). Metabolic profiling of in vitro micropropagation and conventionally greenhouse-grown ginger (*Zingiber officinale*). *Phytochemistry*. 67(20):2239-2255.

Misalova A, urkovic J, Mamonova M, Priwitzer T, Lengyelova A, Hladka D and Lux A. (2009). Changes in leaf organisation, photosynthetic performance and wood formation during ex vitro acclimatisation of black mulberry (*Morus nigra* L.). *Plant Biology*. 11(5):686-693.

Mishra Y, Usmani G and Mandal AK. (2010). In vitro cloning of [*Rauvolfia serpentina* (L.) Benth. var. CIM-Sheel] and evaluation of its field performance. *Journal of Biological Research*. 13:85.

Mohtashami L, Amiri M S, Ayati Z, Ramezani M, Jamialahmadi T, Emami SA and Sahebkar A. (2021). Ethnobotanical uses, phytochemistry and pharmacology of different *Rheum* species (Polygonaceae): a review. *Pharmacological Properties of Plant-Derived Natural Products and Implications for Human Health*. 309-352.

Mokotedi ME, Watt MP and Pammenter NW. (2009). Analysis of differences in field performance of vegetatively and seed-propagated *Eucalyptus* varieties I: survival and leaf gas exchange. *Southern Forests*. 71(4):267-271.

Qin X, Wang Y, Li A, Sun A, Yu L and Liu R. (2017). Separation and purification of six components from the roots of *Rheum officinale* Baill by supercritical fluid chromatography. *Journal of Liquid Chromatography & Related Technologies*. 40(3):156-164.

Romaniello MP. (2016). True rhubarb? Trading Eurasian botanical and medical knowledge in the eighteenth century. *Journal of Global History*. 11(1):3-23.

Shahzad A, Faisal M and Anis M. (2007). Micropropagation through excised root culture of *Clitoria ternatea* and comparison between in vitro–regenerated plants and seedlings. *Annals of Applied Biology*. 150(3):341-349.

Singh R, Tiwari T and Chaturvedi P. (2017). *Rheum emodi* Wall ex. *meissn* (Indian Rhubarb): Highly endangered medicinal herb. *Journal of Medicinal Plants Studies*. 5(4):13-16.

Stoleru V, Munteanu N, Stan T, Ipatioaie C, Cojocaru L and Butnariu M. (2019). Effects of production system on the content of organic acids in Bio rhubarb (*Rheum rhabarbarum* L.). *Romanian Biotechnological Letters*. 24(1):184-192.

Xiang H, Zuo J, Guo F and Dong D. (2020). What we already know about rhubarb: a comprehensive review. *Chinese Medicine*. 15(1):1-22.

Zhao Y, Grout BW and Crisp PC. (2005). Variations in morphology and disease susceptibility of micro-propagated rhubarb (*Rheum rhaponticum*) PC49, compared to conventional plants. *Plant Cell, Tissue and Organ Culture*. 82(3):357-361.

Zhao Y, Grout BW and Crisp PC. (2007). Screening morphological variation in micro-propagated plants of rhubarb (*Rheum rhaponticum*) derived from uniform, acclimated plantlets. In III *International Symposium on Acclimatization and Establishment of Micro-propagated Plants*. 812:521-526.

23

Advances in Production Technology Of Potato

***Chhail Bihari*[*1], *Abdul Rahman M*[2], *Sajeel Ahamad*[3], *Rajat Rajput*[4] *and Ajeet Kumar*[5]**

[1,2,3&5]*Ph.D. Research Scholar, Department of Horticulture, Nagaland University, School of Agricultural Sciences, Medziphema, Nagaland*

[3]*Ph.D. Research Scholar, Division of Food Science & Postharvest Technology, Indian Agricultural Research Institute, New Delhi, India*

[*]*Email: cbrathour500@gmail.com*

Introduction

Potato, scientifically known as *Solanum tuberosum*, is a starchy tuberous crop that belongs to the nightshade family, Solanaceae. It is one of the most widely consumed and important food crops worldwide. Potatoes are native to the Andes region of South America and have been cultivated for thousands of years. They have since become a staple food in many countries and have significant cultural and economic importance. Potatoes are known for their versatility, as they can be prepared and cooked in various ways. They are typically harvested from underground stems called tubers, which come in a variety of shapes, sizes, and colours. The most common varieties have brown or yellowish skin and white, yellow, or creamy flesh. However, there are also red, purple, and even blue-skinned potatoes available. The cultivation of potatoes is widespread, with major potato-producing countries including China, India, Russia, the United States, and Germany. Potatoes are adaptable to different climates and can be grown in diverse environments. They are an important crop for food security and economic stability in many regions. Beyond their culinary significance, potatoes have also contributed to scientific advancements. In the mid-19th century, the Irish Potato Famine caused by a plant disease highlighted the importance of genetic diversity in crop cultivation and led to the development of plant breeding techniques.

Area and Production

Potato (*Solanum tuberosum* L.) is the 3rd most important potential food crop after rice and wheat in the world. With a production and productivity of 56.17

million MT and 24.00 t/ha, respectively, across an area of 2.20 million ha, India comes in second place to China. In India Uttar Pradesh, West Bengal, and Bihar nearly 3/4 of area and 4/5 of potato production in our country. Uttar Pradesh is the leading producer followed by West Bengal, Bihar, Gujrat, and Madhya Pradesh. To fulfil the need of the expanding population and economy, India is predicted to need to produce 122 million MT of potatoes by 2050. It is most important where cultivable grounds area is less. By 2050, when the world population is estimated to increase from the present 6.3 billion to 9.3 billion, 90% of whom would live in Asia, Africa, and Latin America, it is predicted that the world's food consumption will have at least doubled. In India, the population has already surpassed 1.0 billion, and by 2050, it is expected that there will be 1.5 billion people living there, making it the largest country on earth. Potato has grown in importance since it delivers a comparably greater economic output per unit area and time. Due of this, even the in our country small and marginal farmers have started growing potatoes. However, in the hills, plains and plateau regions of our nation, several biotic and abiotic variables repeatedly restrict the potential output of potatoes.

Importance and Uses

It is designated as "*Food for Future*" by FAO due its high nutritive value and productivity (Khan *et al.*, 1994). Nutritionally, potatoes are a rich source of carbohydrates, particularly in the form of starch, which provides energy. They also contain dietary fibre, vitamins (such as vitamin C and vitamin B6), minerals (including potassium and manganese), and antioxidants. The nutritional content of potatoes can vary depending on the variety and cooking method. Potatoes are consumed in numerous ways, including boiling, baking, frying, mashing, and roasting. They are used to make a wide range of dishes, such as french fries, mashed potatoes, potato salad, and potato chips. Additionally, potatoes are used as ingredients in soups, stews, casseroles, and many other recipes. They can be enjoyed as a main course or as a side dish, and their versatility allows them to be incorporated into various cuisines around the world.

It is stapled food of several Western countries. It is a chief crop for high-population areas due to its producing more dry-matter food per unit area as compared to other crops. In the world, approximately 50 per cent of produced potatoes are utilized in human food. In India, about 15 kg of fresh potato tubers are consumed per person per year, as compared to 193 and 126 kg in Denmark and Netherlands respectively. Per capita, the availability of potato tubers has increased from 4.38 kg in 1950 to 26 kg in 2020. Potato is wide scope in production and consumption in the country.

Origin and Distribution

Potato is originated from South America, where it grows wild and exhibits the greatest diversity includes shapes, sizes, colours and tastes for its tubers. The wild relatives of the cultivated potato are said to have lived in the high highlands of Peru and Bolivia near the Tritica basin. Only forty years after the potato's debut in Europe, it was introduced in India Atom Europe around the beginning of the 17th century. Potato is grown all over India except Kerala. Approximately 82 per cent area comes under potato crop in plains area which crop is grown during short-days of winter in October-March whereas only 10% area comes under hills which crop is grown during long-days of summer from April-September.

Botany and Floral Biology

The growth habit of the potato, a herbaceous plant, differs across and within species. It has a tendency to develop rosettes or semi-rosettes. The potato tuber is a stolon, or underground stem, that has grown larger. The buds that will sprout next season's growth are known as tuber eyes. Fewer eyes are located towards the stolon or basal end of the tuber, where eyes are denser. The variation is distinguished by the number and distribution of eyes. Depending on the cultivar, the terminal inflorescence of potatoes can have 1–30 flowers, however, typically 7–15 is present. Flowers are actinomorphic and hypogynous, and the inflorescence is cymose. The arrangement of floral parts is uniform. The flower's five petals are arranged in a star pattern. The form and size of the sepal lobes vary according on the cultivar. Five stamens alternate with five petals to make the androecium (Howard, 1970). The ovary is concealed by a cone-shaped structure made up of all the anthers. Except for male sterile plants, when the anthers are pale yellow or yellow green in colour, anthers are brilliant yellow or orange in colour. Ovules are positioned around the placenta's perimeter in the superior, bilocular ovary. A 12–14-hour photoperiod with a nighttime temperature of 12–15 °C are favorable for flower formation and berry setting. Short days during the flowering season may cause the floral bud to abscise. Multiple factors, including genotype, temperature, photoperiod, inflorescence location, plant/stem density, competition between flower and tuber, precipitation, planting date, and nutritional level, affect the formation of potato flowers and fruits. Flowers stay open for two to four days, and during this time, pollen production and stigma receptivity last for around two days. Berries are the type of fruit, and they have spherical to ovoid shapes with a diameter of around 14 cm. When mature green or green-tinted berries display bands or dots of white or purple.

Taxonomy

The potato is a tuber vegetable crop that is a member of the vast genus *Solanum* and the family Solanaceae. About 235 out of the 2000 species that make up the potato are tuber-bearing (Hawkes, 1990). Ploidy level range of potatoes from 2x to 6x. The domesticated potato is autotetraploid (2n = 4x = 48), and it has two subspecies: *ssp. andigena,* which is adapted to short days, and *ssp. tuberosum,* which is adapted to long days. (X = 12) is the basic chromosomal number. *Solanum stenototum, Solanum goniocalyx, Solanum phureja,* and *Solanum ajanhuiri* are a few examples of diploid species. *Solanum juzepczukii* and *Solanum chaucha* are triploid species. *Solanum curtilobum* is a pentaploid species. Tetraploid and highly heterozygous in nature, potatoes are mainly cultivated. More than 50 traits should be incorporated into the current potato variety (Hawkes, 1994).

Soil and Climate

Generally, potato cultivation is done in all types of soil like (alluvial, red, hill, laterite and black) with a pH range of 5.5-8.0. The best soils are deep alluvial ones in the Indo-Gangetic plains with practically neutral or alkaline soil reactions. Alluvial soils have the most area planted with potatoes, followed by hill, black, and red soils. Soils that are sodic, alkaline, or saline are not good for growing potatoes. Fine, loose soil without compacted layers that prevent root penetration is ideal. Water drainage is also impeded by compacted layers. Stones and clods restrict the amount of soil contact for roots and distort tubers. Potato cultivation is best in soils that are well-drained, coarse or sandy loam to loamy and rich in organic matter. Although the potato is mainly a crop of temperate climates, there is significant genetic variation regarding how the crop reacts to term periods. In India, the potato crop is typically grown when the maximum daytime temperature is below 35°C and the maximum nighttime temperature is not higher than 20°C. When the days are bright and the nights are cool, crops grow well. When the nighttime temperature exceeds 23°C, the majority of potato genotypes do not tuberize and such areas are not to be suitable for potato cultivation. The optimum temperature for potato growth and development ranges from 15-25 ^{0}C. The most suitable temperature for potato tuberization is 17 ^{0}C.

Varieties

Recommended Varieties	Maturity (Days)	Agro-climatic zones
Kufri Ashoka, Kufri Surya, Kufri Jawahar, Kufri, Chandramukhi, Kufri Khyati and Kufri Pukhraj	Early (70-90)	North-Western Plains (Rajasthan, Haryana & Punjab)
Kufri Chipsona-3, Kufri Sutlej, Kufri Gaurav, Kufri Jyoti, Kufri Anand, Kufri Pukhraj, Kufri Pushkar, Kufri Garima, Kufri Sadabahar, Kufri Arun, Kufri Badshah, Kufri Bahar, and Kufri Chipsona.	Medium (90-110)	
Kufri Jawahar, Kufri Chandramukhi, , Kufri Lauvkar, Kufri Surya, Kufri Pukhraj.Kufri Khyati and Kufri Pukhraj	Early (70-90)	North-Central Plains (North-western districts of Western U.P. and M.P.
Kufri Arun, Kufri Chipsona-3 Kufri Jyoti, Kufri Pukhraj, Kufri Badshah, Kufri Frysona, , Kufri Sadabahar, Kufri Sutlej, Kufri Bahar, Kufri Chipsona-1, Kufri Garima, Kufri Gaurav, Kufri Anand and Kufri Pushkar	Medium (90-110)	
Kufri Sindhuri	Late (more than 110)	
Kufri Khyati, Kufri Surya, Kufri Chandramukhi, Kufri Ashoka and Kufri Pukhraj.	Early (70-90)	North-Eastern Plains (Plains of Assam, Bihar, Eastern Uttar Pradesh, Odisha, Northeastern Madhya Pradesh and West Bengal)
Kufri Lalit, Kufri Arun, Kufri Kanchan, Kufri Frysona Kufri Garima Kufri Bahar, Kufri Chipsona-1, Kufri Chipsona-3, Kufri Gaurav, Kufri Jyoti, Kufri Lalima, Kufri Pushkar, Kufri Sutlej, TPS Population 92-PT-2 and Kufri Pukhraj	Medium (90-110)	
Kufri Sindhuri	Late (more than 110)	
Kufri Pukhraj, Kufri Lauvkar Kufri Jawahar, Kufri Khyati, Kufri Surya and Kufri Chandramukhi	Early (70-90 days)	Plateau Region (Odisha, Gujarat, Karnataka, Madhya Pradesh Southern parts of Bihar and Maharashtra)
Kufri Pukhraj, Kufri Garima, Kufri Jyoti and Kufri Badshah	Medium (90-110)	

Kufri Himalini, Kufri Jyoti, Kufri Giriraj, Kufri Girdhari, Kufri Shailja and Kufri Himsona.	Medium (110-120)	North-Western Hills (Himachal Pradesh and Southern Jammu and Kashmir)
Kufri Giriraj, Kufri Himalini, Kufri Jyoti, Kufri Megha, Kufri Shailja and Kufri Girdhari	Medium (110-120)	North-Eastern Hills (Hills of Meghalaya, Mizoram, Nagaland, Arunachal Pradesh, Manipur and Tripura
Kufri Girdhari, Kufri Kanchan and Kufri Jyoti	Medium (110-120)	Sikkim Hills (North Bengal hills and Sikkim) and North Bengal
Kufri Jyoti, Kufri Girdhari, Kufri Neelima, Kufri Shailja, Kufri Giriraj, Kufri Himalini and Kufri Swarna.	Medium (110-120)	South Indian Hills (Hills of Tamil Nadu)

Important Varieties and Their Features

Kufri Manik: It is rich in micronutrients (Fe and Zn), anthocyanin, and carotenoids and suitable for eastern plains (bio-fortified variety). It is resistant to late blight disease.

Kufri Neelkanth: It is resistant to late blight and rich in antioxidants (anthocyanin and carotenoids), excellent flavour and specialty potato.

Kufri Lima: It is susceptible to late blight and extremely resistant to PVX and PVY viruses. And tolerant to early heat, hopper burn and mite and suitable for early and main planting.

Kufri Sahyadri: This is high resistant to potato cyst nematodes & moderately resistant to late blight.

Kufri Sangam: It is having excellent storability and moderately resistant to late blight.

Kufri Thar-1: It is drought tolerant (20% water saving), and mainly suitable for Orissa & Uttar Pradesh.

Kufri Thar-2: It is drought tolerant (20% water saving), and suitable for Uttar Pradesh, Rajasthan, Haryana, and Chhattisgarh.

Kufri Thar-3: Drought tolerant (20% water saving), and suitable for Haryana, UP and Chhattisgarh.

Kufri Karan: High resistant to late blight, six potato viruses and potato cyst nematodes

Kufri Girdhari: Highly resistant to late blight and long tuber dormancy

Kufri FryoM: Resistant to late blight and PVY, and suitable for French fries and processing

Immune to wart and resistant to late blight: Kufri Jyoti, Kufri Kanchan and Kufri Sherpa **Resistance to frost:** Kufri Sheetman, Kufri Chipsona-2

Resistant to potato cyst nematode and late blight: Kufri Swarna and Kufri Thenmalai

Resistant to golden cyst nematode: Kufri Swarna and Kufri Dewa

Resistant to late blight, early blight, potato virus X: Kufri Badshah

Resistant to late blight: Kufri Jyothi, Kufri Giriraj

Suitable for processing of light colour chips and finger fries: Kufri Chipsona-1, 2 and Kufri Frysona

Propagation

Tubers are the conventional methods of potato propagation. Axillary buds are found in the eyes on the surface of tubers. Dormancy of potato tubers having around 8 to 10 weeks. The axillary buds begin germination and sprouting after over the dormancy. When these sprouting tubers are placed in the soil, they develop quickly and vigorously. Plant density, which is dependent on the number of main stems, has a significant impact on the production of potatoes and the number of tubers. The multiple sprouting stage is the best stage for planting, due to it generating the emergence of several strong stems. Therefore, pre-sprouting of seed stocks aids in raising the yield and, as a result, the number of main stems. A speedy, uniform, and complete germination is also ensured. The seed stocks are removed from cold storage 10-15 days before to planting and held in pre-cooling chambers for 24 hours. They are then spread out in a cool and shaded place for sprouting. The only seed tubers are planted which sprouted nicely.

Planting

Planting In different locations of the country, various techniques are utilized to plant potatoes after pre-planting tillage activities. The most traditional strategy, whether carried out manually or automatically, is the ridge and furrow method. Using a curved or narrow-bladed spade, the furrows are made in the manual method, which is followed by the application of a fertilizer combination, the covering of the soil, and the make of ridges. In the mechanical method, furrows are produced with the aid of tractor-drawn 2-4 row marker cum fertilizer drills to apply fertilizer in one sequence, whilst the seed tubers are excavated on the North side of each ndge. When there is no mechanical planter or fertilizers drill, ridges are made using a tractor-drawn ndger after fertilizers application, and tubers are manually planted 5-7 cm deep on the ridges. Another way involves marking the field with a rope or marker and then applying fertiliser to the lines that have been drawn. On top of these lines, tubers are put, and ridges are made manually or with tools pulled by bullocks. In the hills, tubers are put and then covered with earth to make ridges after fertilizers has been applied in shallow furrows drawn with hand tools (Khiana or Kudat). All of these methods take precautions to prevent seed tubers from coming into direct touch with fertilizer.

Manure and fertilizers

Since it is a tuber vegetable, more nitrogen is requiring for crop growth and development. The Standard dosage is 20 -25 tones FYM and 120:80:80 kg/ha of N, P and K are given. At the time of field preparation full does of FYM should be applied and incorporated into soil. The full dose of P and K and half does of N should be applied. After every remaining dose of nitrogen should be apply via top dressing at the time of earthing up (45 DAP). Potato is exhaust high amount of K followed by N and P (Saini *et al.*, 2021).

True Potato Seed (TPS)

Two methods were followed for the seed production of potatoes through TPS which describe below (Almekinders, 1991).

1. Transplanted seedling method

TPS is initially sown in a nursery to raise seedlings before transferred onto a properly prepared seed bed. This process is very labor- and money expensive (Sharma and Dhakar, 2013).

2. Seedling tubers method

The true potato seed was tested and tried at International Potato Centre (CIP) in lima Peru and Central Potato Research Centre Institute (CPRI) in Shimla. The best strategy to avoid raising seedlings before transplanting is to use this method (Almekinders, 1996).

TPS	
Seedling tubers method	**Transplanted seedlings method**
1. Seedling tuber production in the nursery • 50 g TPS seeds and 375 m² area are enough for production of seedling tubbers for planting one ha field. • 2-3 seeds are sown @ 10 x 10 cm spacing at 0.5 cm depth. • After attaining seedling height of about 15 cm, cover the lower most 3 internodes with additional subtract mix. • Haulmcutting is done before about 15 days harvest the seedling tubers. • Four grades of the tubers which are grading as < 5g , 5-10 g,10-20 g and > 20g • 3% boric acid solution taken for Soak seeds for 30 minutes after that seed dry and store for using as seed in the season. • **2. Seedling tuber production in the field** • Prepare the nursery beds and carry out all procedures in accordance with the preceding approach, except that of increasing the row-to-row gap to 30 cm to allow earthing up of the plants.	**1. Preparation of nursery beds** • For planting 1 ha field 150 g TPS and 75 m² area of nursery are required. **2. Seedling raising** • TPS sown at the depth of ½ cm. • Seed beds Irrigate 3 times in a day • 20-25 days old seedling ready for Transplanting **3. Field Preparations** • To prepare the field, use a traditional technique by creating ridges that are 20 cm high and 45–50 cm apart in an east–west direction.
3. Potato production in the field • Sow your crop using the seedling tubers that were obtained in the previous crop season. • Spacing maintained for mechanical 60 x 10 cm and for manual cultivation 45-50 x 10 cm. • When cultivating a crop using seed tubers, all cultural and traditional practices are followed.	**4. Seedling transplanting** • For transplant the seedling spacing should be @ 45-50 x 10 cm. • transplanting and Irrigation are sequential operations. • Earthing up do at 30-35 Days after transplanting • All subsequent cultural operations do and take plant protection measures as per recommended cultivation in that region.
4. Adoptability seedlings tubers are produced in the first year. These tubers and tuberlets are planted to grow the commercial crop the next year (Struik, 2007).	**5. Adoptability** • State like Assam, Bihar, Eastern UP, Gujarat, etc. adoptability is mild winter areas. • Using TPS Commercial crop can be raised in the first year.
5. Yield: The yield is higher.	**6. Yield:** The yield is comparatively less.
Recommended varieties of TPS: H.P.S. 1/13, T.P.S.C. -3, 92 PT 27	

Intercultural Operations

For consistent germination, watering before planting is beneficial. After about a week, a second irrigation is administered, and more are provided as needed. Irrigations that are light and frequent are preferable than those that are heavy and infrequent. At crucial junctures in crop development, water is applied efficiently and economically. stages of the crop: stolon creation, tuber initiation, and tuber development. In general, potato needs 700 mm of irrigation water for the optimum production. About ten days before crop harvest, irrigation is discontinued to allow the skin of the tubers to firm up. Mulching promotes fast germination, lowers soil temperature, and preserves soil moisture. Materials that are readily available locally, such pine needles or leaf litter, are very good at reducing runoff losses and preserving moisture. Weeds significantly reduce productivity by competing with crops for nutrients, moisture, light, and space. Chemical, cultural, or a combination of the two strategies are effective ways to control weeds. When the crop is about a month old, weeds can be controlled by hoeing, followed by earthing up. Fluchloralin @ 1 kg a.i per ha, alachlor @ kg a per ha, pendimethalin @ 1.8 kg a.i per ha, or atrazine @ 1.0 kg a.i per ha can all be applied pre-emergence to effectively control the weeds. Paraquat @ 0.36 kg a.i. per hectare applied post-emerging (only 5-10% plant emergence) is also effective. To manage the weeds, apply Toke-25 @ 2.5 kg ai per ha as a post-emergence application at about 2-3 leaf stage.

Physiological Disorder and Their Management

1. Hollow Heart

It is due to the fast growth of tubers formation. Tubers become large and remain empty with in leading to the formation of cavity in the center with the death of the small area of pith cells. This results in adjacent cracks and hollowness as the center expands during the growth of the potato. It is overcome by maintain soil moisture conditions and avoid overdose of fertilizers particularly Nitrogen. And used resistant varieties.

2. Black Heart

Black heart is another disorder which caused by sub-oxidation conditions under potato tuber storage in heaps as the air does not reach into the center. It is mostly due to higher temperature and excessive moisture resulted in blackening of tissues in the center. It is overcome by proper ventilation and keep the potato tubers in layers. Do not store tubers in the heap.

3. Greening

There are many factors that increase the glycoalkaloid content of potato tubers. Mechanical damage, early harvest, excessive fertilizer use, or overexposure to sunshine can all produce glycoalkaloid, which increases the formation of the somewhat toxic solanine. As tuberization occurs and after tubers are digging up, they are stored in darkness, this problem is solved by properly earthing the tubers.

4. Knobbiness

It is due to uneven watering conditions and unequal tuber cell/tissue growth are to blame for the restriction in tuber growth. After a prolonged dry season, high irrigation causes some cells to develop quickly, resulting in knobbiness. It is overcome by optimum irrigation supply is the remedy.

5. Tuber cracking

Boron deficiency or uneven water supply is responsible for cracking. It is overcome by the use of Borax @ 20 kg/ha. It is overcome by frequent and optimum irrigation supply is the good remedy.

6. Sun Scalding

Typically, it happens around the time of the autumn harvest when the weather is warm and there is more sunshine. At that point, tip burn has a significant impact on the emergence of sprout and leaflets. When the temperature exceeds 30°C, it emerges. The soil temperature is controlled by passing water through the furrows.

7. Translucent End

Heat stress and drought are responsible for it. Usually, the proximal end of the tuber is where it may be located. Potato tubers have an uneven form and a shiny sheen. In addition, this causes storage-related deterioration. These shiny regions contain little total soluble solids and a lot of sugar. It is managed by Maintain 50% moisture in the field and avoid excessive nitrogen supply.

8. Black Spot

Within three days of mechanical damage, vascular tissues in potato tubers begin to discolor internally. The black spots on potato tubers is related to phenols. It is overcome by provide proper storage and growing conditions and genetic make-up of the varieties.

9. Freezing Injury

It is a result of tubers being exposed to cold temperatures during or right after harvest. It occurs at a temperature of -1.5°C or lower. Net necrosis refers to the loss of small vascular ring components or cells. Ring necrosis is a tissue darkening that affects the vascular tissues there. This results in tubers that can't be sold. The proximal end of the tubers exhibits higher damage. Avoiding exposing tubers to cold temperatures during harvest or storage helps to overcome it.

10. Sprouting

It is a very serious problem in storage. Spraying MH at 1000–6000 ppm around 2-3 weeks prior to harvest is effective in overcoming it. Additionally, additional substances like Chloro IPC (N-tetra chloro isopropyl carbonate) at 0.5% and/or nomyl/amyl alcohol at 0.05–0.12 mg/ha aid in preventing potato tuber sprouting.

11. Swollen Lenticels

This disorder is caused by the tubers being exposed to very humid circumstances in the field or during storage due to a lack of oxygen. The primary issue is the creation of an entry point for pathogenic organisms, bacterial soft rot, pink rot, and leak. It is overcome by the avoid over-watering. Pick fields with good drainage.

Plant protection

1. Early blight

It is caused by *Alternaria solani*. The typical characteristic of early blight is spots with concentric rings of brown to black colour are formed on the leaves. When heavily infection is occurred then leaves fall off after drying. Spots is also appeared on stems.

Management

It is managed by destroy the crop debris through burning. Chemical control like to spray mancozeb or zineb @ 0.2% or 0.3% copper oxychloride at fortnight interval 3-4 times. Grow resistant varieties like Kufri Jeevan and Kufri Naveen.

2. Late blight

It is very serious disease of potato caused by *Phytophthora infestans*. The lower leaves are infected generally from margin or apex and having cottony growth on the lower side. Water-soaked lesions appear on the margins. Tuber's rotten under very severe infection.

Management

The use of disease-free certified seed and follow crop sanitation. Spray crop with ridomil MZ 72 @ 0.2% (2g/lit of water) at 15 days interval from first appearance of disease. Grow resistant varieties like Kufri Himsona and Kufri Giriraj.

3. Black scurf

It is caused by *Rhizoctonia solani*. Cankers cause wilting of plants and sprouts are killed before emergence. Black crust on potato tubers gives ugly appearance.

Management

Use disease free certified seed. Seed tubers should be Treated with recommended fungicides. And follow the crop rotation which is also effective.

4. Brown rot

It is a serious disease caused by *Ralstonia solanacearum*, wilting and stunting of plants occur. The brown rot is a soil and seed borne disease.

Management

It is overcome by the use of disease-free tubers and follow crop rotation with cereal crops like wheat and maize.

5. Root knot nematode

Keep land fallow for a quite long time. Follow crop rotation. Nematicides like DD @ 225 l/ha should be injected in the soil. Place between the rows EDB @ 175 kg/ha 2 weeks before planting.

1. Cut worm (*Agrotis ipsilon*)

Dome shaped eggs with creamy white colour, laid singly on lower surface of the leaves. Dark brown pupae are found in earthen cells lying underground in the potato fields. The newly emerged young larva is yellow in colour. Adult moth is dark with some grayish patches on the back and dark streaks on the forewings. Young larvae feed on the epidermis of the leaves. Older larvae come out at night and feed young plants by cutting their stems. They also damage the tubers by eating away part of them. They feed tender part only.

Management

It is managed by flood the infested fields, hand picks and destroys the larvae during morning and evening hours on cracks and crevices in the field. Field plough during summer months to expose larvae and pupae to avian prodators. Set up light trap @ 1/ha. Pheromone traps @ 10/ha to stract male moths. Spray chemical like chlorpyriphos 20EC @ Iml/1 or neem oil @ 5 ml/ 1 Apply phorate 10G granules @ 2 kg/ha on soil around the plant.

2. Potato tuber moth (*Phthorimaca operculella*):

Eggs are laid singly on the ventral surface of foliage and exposed tubers. Larva is yellow coloured and caterpillar has dark brown head. Pupation occurs within a cocoon among de trash and clods of the earth in the field. Adult is small narrow winged moth with brown forewings and hind wings are dirty white. It is a pest of dd and storage. Larva tunnels into foliage, stem and tubers. Galleries are formed near tuber eyes.

Management

Select healthy tubers and avoid shallow planting of tubers. Plant the tubers at depth of 10-15 cm. Install pheromone traps @15 numbers/ha. Collect and destroy all the infested tubers from the field Do earthing up at 60 days after planting to avoid egg laying on the exposed tubers. Spray NSKE 5% or quinalphos 25 EC 60 2ml/1 of water to manage foliar damage. Spray Bacillus thuringiensis kg/ha at 10 days interval.

3. Aphids (*Mycus persicae*)

They are found lower surface of leaves and suck the sap from leaves. It is work like vector and spread the diseases like potato mosaic virus.

Management

Apply phorate 10G granules @ 2 kg/ha on soil around the plant. Spray the crop with rogor or metasystox (@ 0.1% and repeat the spray at 10-12 days interval.

Harvesting and Yield

Before the temperature rises over 30°C, potatoes are harvested. To prevent tuber rot caused on by high temperatures in March/April, it is completed by the end of January in the central and eastern plains and by February 15 in the western plains. 10-15 days after the last irrigation is stopped, then crop is harvested. As a result, tuber skin might become hard and avoid bruising during harvest. Using a spade or khurpv, or a bullock-drawn single row digger/plough, harvesting is done manually. Additionally, it can be carried out mechanically using a 1-4 row potato digger. Early varieties 200 quintal per hectare and late varieties 300 quintal per hectare.

Post-Harvest Handling, Grading and Storage

About a fifth of the nation's entire potato harvest is utilized as planting the subsequent season. Post-harvest treatment, especially regarding seed stocks, is therefore important. Potatoes are stored in heaps in cool areas for an additional 10 to 15 days after harvesting in order to dry and cure the skin. The ideal heaps are 3–4 meters long, 1 metre wide at the base. In the hills' harvested potatoes are spread out for drying in well-ventilated chambers. All cut, damaged and rotten tubers are removed prior to grading (Spooner and Salas, 2006). After that potato tubers are graded according to size and put in gunny bags. Generally, the four grades of potato are, small (below 25g), medium (25–50g), large (50–75g) and extra-large (above 75g). After grading, potatoes planned for seed usage the following year are treated with a 3% boric acid solution for 30 minutes to protect against soil-borne diseases, such as common scab and black scurf, before being stored in bags.

After drying, curing, and grading, seed potatoes are kept in cold warehouses in the plains where the temperature is kept between 2- 4 degrees Celsius with high relative humidity. Low temperatures prevent sprouting and decay, while high relative humidity reduces tuber weight loss.

Future Prospect

The future prospects for crop improvement of potatoes are promising. Researchers are focusing on enhancing disease resistance to combat common potato diseases such as late blight and viruses. They are also striving to increase yield potential through improved photosynthetic efficiency and tolerance to environmental stresses. Efforts are being made to enhance the nutritional value of potatoes by increasing beneficial compounds and resistant starch content. Additionally, there is a focus on developing varieties with improved storage and processing characteristics, such as longer shelf life and better texture after cooking. Overall, advancements in breeding techniques and genetic engineering hold great potential for developing improved potato varieties that are disease-resistant, higher-yielding, and more nutritious.

References

Almekinders, C. J. (1991). Flowering and true seed production in potato (*Solanum tuberosum* L.). 2. Effects of stem density and pruning of lateral stems. *Potato research*, *34*, 379-388.

Almekinders, C. J. (1992). The effect of photoperiod on flowering and TPS production in the warm tropics. *Potato Research*, *35*, 433-442.

Almekinders, C. J. and Wiersema, S. G. (1991). Flowering and true seed production in potato (*Solanum tuberosum* L.). Effects of inflorescence position, nitrogen treatment, and harvest date of berries. *Potato research*, *34*, 365-377.

Almekinders, C. J. M. and Struik, P. C. (1996). Shoot development and flowering in potato (*Solanum tuberosum* L.). *Potato Research*, *39*, 581-607.

Chadha, K.L. (2019). Hand Book of Horticulture. Second Edition, Volume 1, Indian Council of Agricultural Research, New Delhi. pp. 592-599.

Gopal, J and MS Rana. (1988). Induction of flowering in potato in North-western plains of India. *J. Indian Potato Asoc.* 15:91-93.

Gopal, J. (2006) Considerations for successful breeding. In: Gopal J, Khurana SMP (eds) *Handbook of potato production, improvement and postharvest management*, Haworth's Press, New York, pp 77-108

Hawkes, J.G. (1990). *The Potato, Evolution, Biodiversity and Genetic Resources*. Belhaven Press, London, 259p.

Hawkes, J.G. (1994). Origin of cultivated potatoes and species relationships. In*: Potato Genetics. (Eds: Bradshaw, J.E. and Mackay, G.R.). CAB International, U.K., pp.3-42.*

Haynes, K.G. (2001). Variance components for yield and specific gravity in a diploid potato population after two cycles of recurrent selection*. Amer. J. of Potato Res*. 78: 69-75.

Howard, H.W. (1970). Genetics of Potato *(Solanum tuberosum L.). Logos Press Ltd., London, pp.20-24.*

Jähnl, G., Zeller, A. and Jähnl, G. (1954). Zur Kenntnis des Blühverhaltens der Kartoffel. *Veröffentlichungen der Bundesanstalt für Alpine Landwirtschaft in Admont*, 40-74.

Khan, I. A., Luthra, S. K. and Ezekiel, R. (1994). Flowering induction in potato by extended photoperiod and application of growth regulators. *Potato: Present & Future*, 46-48.

Narayan, S., Kanth, R. H., Khan, F. A, Narayan, R. and Chatto, M. A. 2016. Effect of accumulated heat units on growth, yield and tuber quality of potato under conditions of Kashmir *The Bioscan An International Quarterly Journal of Life Sciences* 11(2):1215-1218

Saini, I., Kaushik, P., Al-Huqail, A. A., Khan, F. and Siddiqui, M. H. (2021). Effect of the diverse combinations of useful microbes and chemical fertilizers on important traits of potato. *Saudi Journal of Biological Sciences*, *28*(5), 2641-2648.

Sharma and Dhakar (2013). Production Technology of Potato Using True Potato Seed. *Popular Kheti.*1(2):71-74.

Sleper, D. A. and Poehlman, J. M. (2006). *Breeding field crops* (No. Ed. 5). Blackwell publishing.

Spooner, D. M. and Knapp, S. (2013). *Solanum stipuloideum* Rusby, the correct name for *Solanum circaeifolium* Bitter. *American Journal of Potato Research*, 90, 301-305.

Acquaah, G. (2009). Principles of plant genetics and breeding. John Wiley & Sons.

Spooner, D. M. and Salas, A. (2006). Structure, biosystematics, and genetic resources. In *Handbook of potato production, improvement, and postharvest management* (pp. 1-40). CRC Press.

Struik, P. C. (2007). The canon of potato science: 40. Physiological age of seed tubers. *Potato Research*, *50*(3-4), 375.

Thijn, G. A. (1954). Observations on flower induction with potatoes. *Euphytica*, *3*(1), 28-34.

Tiwari, J. K., Luthra, S. K., Bhardwaj, V., Singh, R. K., Buckseth, T., Kumar, R. and Kumar, M. (2021) Indian Potato Varieties. *Indian Horticulture*, 64(1).